Chemistry of

Carotenoid

Radicals and Complexes

Other Titles by Lowell D. Kispert

Electron Spin Double Resonance Spectroscopy

Chemistry of Carotenoid Radicals and Complexes

Lowell D Kispert
University of Alabama, Tuscaloosa, USA

A Ligia Focsan
Valdosta State University, GA, USA

NEW JERSEY • LONDON • SINGAPORE • BEIJING • SHANGHAI • HONG KONG • TAIPEI • CHENNAI • TOKYO

Published by

World Scientific Publishing Co. Pte. Ltd.
5 Toh Tuck Link, Singapore 596224
USA office: 27 Warren Street, Suite 401-402, Hackensack, NJ 07601
UK office: 57 Shelton Street, Covent Garden, London WC2H 9HE

Library of Congress Cataloging-in-Publication Data
Names: Kispert, Lowell D., 1940– author. | Focsan, A. Ligia, author.
Title: Chemistry of carotenoid radicals and complexes / Lowell D. Kispert, University of Alabama, USA, A. Ligia Focsan, Valdosta State University, USA.
Description: Hackensack, NJ : World Scientific Publishing Co. Pte Ltd, [2024] | Includes bibliographical references and index.
Identifiers: LCCN 2023026150 | ISBN 9789811278341 (hardcover) | ISBN 9789811278358 (ebook fo institutions) | ISBN 9789811278365 (ebook other individuals)
Subjects: LCSH: Carotenoids. | Carotenoids--chemistry.
Classification: LCC QP671.C35 K47 2024 | DDC 612/.01528--dc23/eng/20230922
LC record available at https://lccn.loc.gov/2023026150

British Library Cataloguing-in-Publication Data
A catalogue record for this book is available from the British Library.

For any available supplementary material, please visit
https://www.worldscientific.com/worldscibooks/10.1142/13471#t=suppl

Desk Editor: Shaun Tan Yi Jie

Typeset by Stallion Press
Email: enquiries@stallionpress.com

Preface

In 1984, using the EPR/ENDOR instrument facility at the Argonne National Lab located in the Chicago, IL area for the purpose of determining the structure of the radicals produced by X-ray irradiation in organic single crystals, a discussion was held by the head of the photosynthesis research group, James Norris, as to the role of carotenoids in plant photosynthesis. It was believed by this group of researchers studying plant photosynthesis at the National Lab that carotenoids only served as space holders to keep the chlorophylls in place for sun absorption to allow photosynthesis to occur, and to prevent the chlorophylls from aggregation during photosynthesis. James Norris stated that it had to be more complex. Investigator Kispert took that as a challenge to propose to the U.S. Department of Energy in 1985 to study the role of carotenoids in plant photosynthesis. Their acceptance resulted in 30 years of carotenoid research funding for 13 PhD students and support for numerous research scholars that produced 121 published papers, examining approximately 15 of the more commonly commercially available carotenoids. It turned out James Norris was correct. Besides stabilizing the pigment-protein complexes, carotenoids are active in harvesting sunlight and in photoprotection and contribute along with chlorophylls to the overall mechanism of energy transport and conversion. Science has proven that carotenoids can no longer be considered as only accessory pigments to chlorophyll.

The physisorption, electron and proton transfer processes that occur when carotenoids are adsorbed on solid artificial matrices or dissolved in aqueous solution have been elucidated in our group, and similar reactions have been predicted in plants. Our results *in vitro* on solid surfaces relevant to photoprotection support not only the formation of radical cation $Car^{\bullet+}$, first detected *in vivo* in Prof. Graham Fleming's group in 2005, but also the formation of the neutral radicals $^{\#}Car^{\bullet}$ by proton loss from $Car^{\bullet+}$. But do neutral radicals occur *in vivo*? If they do, they could provide another effective non-photochemical quencher of the singlet and triplet excited states of chlorophyll. Therefore, we urge the scientific community to consider proton loss from $Car^{\bullet+}$ and the possible formation of these radical species *in vivo*.

We have provided here instructive procedures for various measurement techniques on carotenoid radicals such as electrochemical techniques in Chapter 2, density functional theory calculations in Chapter 3, electron paramagnetic resonance techniques in Chapter 4, a discussion of carotenoid complexes in Chapter 5 and photoprotection-related studies in Chapter 6. We conclude in Chapter 7 suggesting different analysis sources available to further study carotenoids. Our goal is for others to be encouraged to take up the challenge to study, examine and characterize the possibly endless number of carotenoids of which approximately 1,200 are known to be remaining.

Lowell D. Kispert, Professor of Chemistry, Emeritus
The University of Alabama, Tuscaloosa, AL, USA

A. Ligia Focsan, Professor of Chemistry
Valdosta State University, GA, USA

About the Authors

Dr. Lowell Kispert is Professor Emeritus at the Department of Chemistry & Biochemistry, The University of Alabama, Tuscaloosa, USA. He is interested in understanding the energy transfer between carotenoids and chlorophyll, as well as the structure of carotenoid excited states, radical cations and triplet states. He holds a PhD in Chemistry from Michigan State University, USA.

Dr. Alexandrina Ligia Focsan is Professor of Chemistry at Valdosta State University, GA, USA. She was elected Fellow of the International Carotenoid Society in 2023. For the past 20 years, she has investigated carotenoid radicals placed in a variety of chemical, material and biological contexts using advanced electron paramagnetic resonance methods, density functional theory calculations, electrochemical measurements, optical studies, etc. She holds a PhD in Physical Chemistry from The University of Alabama, Tuscaloosa, Professor Kispert being her graduate advisor. Over the next decades, Prof. Kispert became her mentor and research collaborator.

Contents

1

Introduction

Carotenoids are naturally occurring, intensely colored pigments formed in numerous photosynthetic bacteria, some species of archaea and fungi, algae, and plants. These intensely colored molecules are responsible for the sources of the yellow, orange, and red colors in many organisms including animals that introduce them in their diet. Animals are incapable of synthesizing carotenoids *de novo* so those found in animals are either directly accumulated from natural food or the feed provided, or partly modified through metabolic reactions. For some species the colors given by unaltered dietary carotenoids or by carotenoid pigments metabolically derived from dietary pigments may be reflected in their appearance. Carotenoids are distributed in animal skins or feathers, and in certain marine organisms such as shrimp, salmon, lobster, or fish egg. As an interesting example, the color of flamingoes' feathers can range from pale pink to intense crimson depending on the amount of carotenoid pigment present in the birds' diet. Flamingoes get their color from eating small brine shrimp, salmon, or crustaceans which in turn get their color from eating algae that synthesize some specific marine carotenoids.

Carotenoids are essential pigments in photosynthetic organs along with chlorophylls. Carotenoids also give the yellow, orange, and red colors to many fruits and flowers but in plants with an abundance of chlorophyll such as green leafy vegetables their color is masked. The leaves in the fall appear yellow, orange or red because as the chlorophyll breaks down, the green disappears, and the colors given by carotenoids become visible. According to the Carotenoids Database established in 2017, currently there are 1,204 naturally occurring carotenoids that have been identified in 722 source organisms from different domains of life: archaeal carotenoids, bacterial carotenoids, and carotenoids in eukaryotes.[1]

Carotenoids have important biological roles in the functioning and survival of plants, such as light harvesting and photoprotection in the process of photosynthesis. Different isomeric forms of carotenoids appear to have different physiological or metabolic roles and play different roles in plant photosynthesis. They are also important dietary nutrients having antioxidant potential. Carotenoids possess free radical-quenching properties. They scavenge singlet molecular oxygen ($^1O_2$) and peroxyl radicals generated in the process of lipid peroxidation.

Fruits and vegetables provide most of the 40 to 50 carotenoids found in the human diet. Fruits and vegetables that are rich in dietary carotenoids include carrot, sweet potato, tomato, bell pepper, broccoli, watermelon, cantaloupe, mango, orange, squash, etc. Dark green leafy vegetables such as spinach, kale, turnip and mustard green are also an important source of carotenoids. Fruit and vegetable consumption is associated with a reduced risk of certain diseases. The optimal intake levels of fruits and vegetables for maintaining long-term health are uncertain. According to Wang *et al.*,[2] 66,719 women and 42,016 men, age 40 and above, were followed on dietary information biennially for 30 years using questionnaires. They concluded that intake of approximately 5 servings per day of fruit and vegetables, or 2 servings of fruit and 3 servings of vegetables, was associated with the lowest mortality, and above this level, higher intake was not associated with additional risk reduction. For a healthy diet, and to reduce the risk of developing conditions such as cancer and cardiovascular disease, the World Health Organization (WHO) also proposes a minimal quantity of 400 grams (5 servings) of fruits and vegetables per day[3] excluding potato, sweet potato, cassava, and other starchy roots.

Carotenoids, which are well-known antioxidants, have been the subject of extensive research due to their potential health benefits. Their antioxidant properties can be beneficial in preventing oxidative stress and the development of certain diseases. There is evidence that carotenoids produce improvements in cognitive function and heart disease, contribute to eye health, and may help to protect against some types of cancers.[4] β-carotene has a pro-vitamin A function, and vitamin A is needed for good vision and eye health, for a strong immune system, and for healthy skin and mucous membranes. Two dietary carotenoids, lutein and zeaxanthin, and the isomer *meso*-zeaxanthin, are found in high concentrations in the macula, the part of the eye responsible for central vision. They are known for contributing to eye health by protecting the macula from the harmful effects of ultraviolet (UV) radiation, such as absorption of blue light or UV-induced peroxidation, and formation of damaging free radicals. A diet rich in foods containing zeaxanthin and other carotenoids, such as leafy green vegetables and eggs, may reduce progression of age-related macular eye disease and cataract. Also, zeaxanthin and lutein are commonly used as dietary supplement to improve the macular pigmentation and visual acuity. Increased lutein and zeaxanthin intake (particularly at doses >10 mg/d) can help with maintaining ocular health.[5] Carotenoids, particularly those found in the eyes and skin, have been shown to provide photoprotection against the harmful effects of UV radiation. Similarly, carotenoids found in human skin (lycopene, α-, β-, γ-, and δ-carotene, β-cryptoxanthin, lutein, and zeaxanthin) protect the skin against sunlight-induced oxidation effects.[6] Lutein was found to have beneficial effects also on the brain, where it was associated with improved cognitive performance.[7] There is emerging evidence suggesting that this could be expanded to include research in Alzheimer's disease, cognitive function, diabetes and cancer.[4]

Despite the evidence for the health benefits of carotenoids, more research, including clinical studies, is needed to confirm these findings and to understand the optimal dietary intake of these carotenoids. In general, there are no univocal data concerning the most appropriate dosage for daily supplementation. Intake recommendation would help to generate awareness in the general population to have an adequate intake of carotenoid-rich foods.[4,7] Organizations such as the International Carotenoid Society play a special role in advancing interest in carotenoids and their beneficial role in maintaining health, but a joint approach by all stakeholders is required to advocate for more research and applied science in humans and animals.[4]

The study of carotenoids for health benefits has endless potential. Scientists examine them in the attempt to find cures to medical problems. For example, in the most recent pandemic, scientists thought of exploring carotenoids as potential new antivirals against SARS-CoV-2.[8] Apocarotenoids like bixin and β-apo-8′carotenoic acid and an ester of crocetin were used for the lipophilic modification of glycopeptide as potential new antivirals against SARS-CoV-2. Two of these naturally occurring carotenoids with carboxyl groups, bixin and β-apo-8′carotenoic acid, exerted remarkable anti-SARS-CoV-2 activity. Bixin, extracted from *Bixa orellana,* is a natural colorant and used in the food industry. Crocetin can be extracted from saffron, which is also used as a food additive. Saffron is traditionally used in cold, fever, bronchitis, respiratory disorders and its anti-inflammatory, antioxidant and immunomodulatory effects are well known.[8]

Figure 1.1 contains chemical structures of the six most prevalent carotenoids. These carotenoids found in diet, blood and tissue are lycopene, α-carotene, β-carotene, β-cryptoxanthin, zeaxanthin and lutein. They are the only carotenoids present in human blood plasma.[9–11] According to their chemical structure, carotenoids can be divided into carotenes

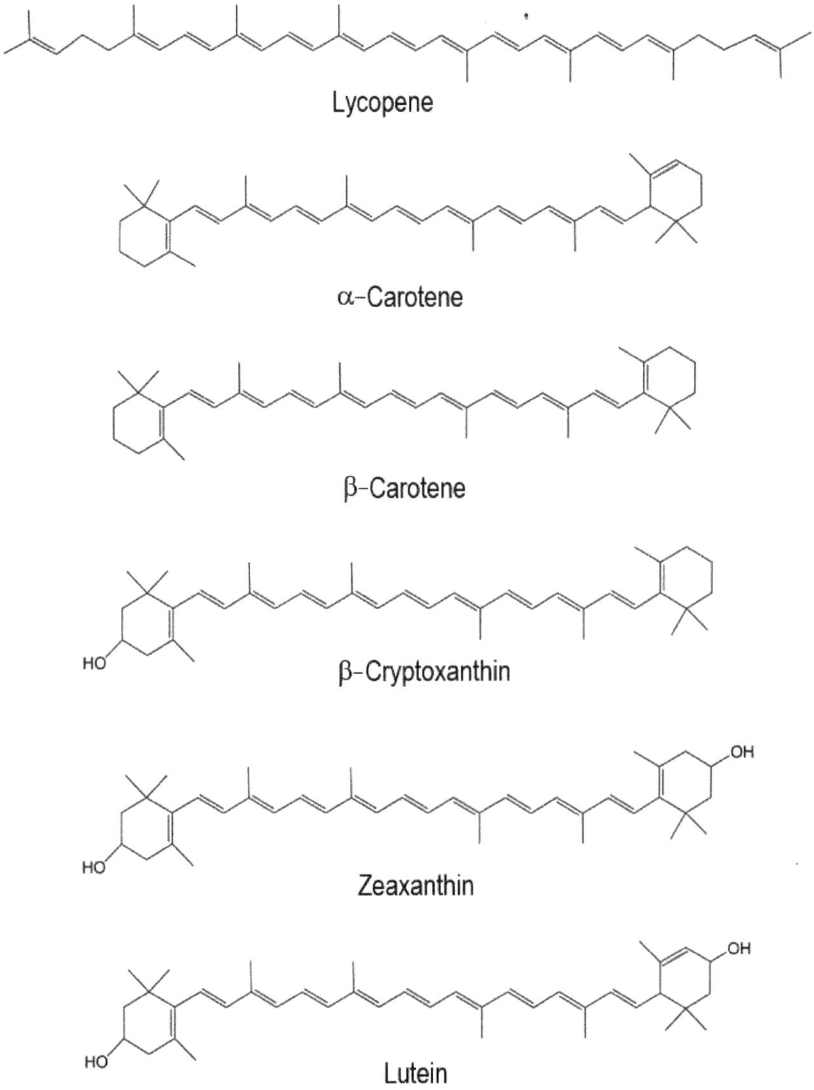

Lycopene

α-Carotene

β-Carotene

HO

β-Cryptoxanthin

OH

HO

Zeaxanthin

OH

HO

Lutein

Figure 1.1. Chemical structures of the six most prevalent carotenoids.

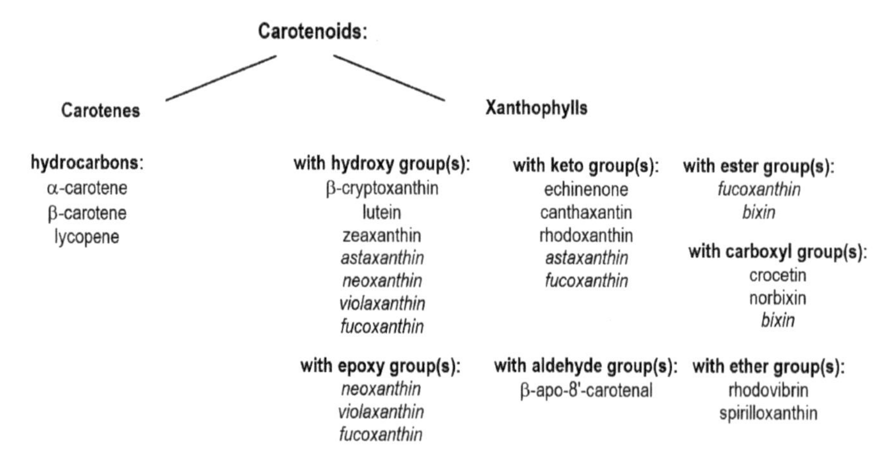

Figure 1.2. Classification of naturally occurring carotenoids on the basis of their chemical nature. Carotenoids in Italic font have more than one functional group.

and xanthophylls. Carotenes are hydrocarbon carotenoids (for example, the first three carotenoids in Figure 1.1), and xanthophylls contain oxygen groups such as hydroxyl (for example, the last three carotenoids in Figure 1.1), carbonyl, carboxyl, epoxy, methoxy, acetate or lactone. Figure 1.2 contains examples of naturally occurring carotenoids and the type of their functional groups. Structurally, carotenoids are long-chain polyenes containing a variable number of conjugated double bonds that constitute the chromophoric group that absorbs in the visual range of the spectrum resulting in their characteristic coloration in yellow, orange, or red. This polyene chain is rich in electrons, and thus susceptible to electrophilic attack by free radicals or oxidizing agents. Carotenoids in solution, *in vitro*, are easily degraded in the presence of just traces of oxygen; however, in *vivo*, carotenoids may be stabilized by lipids or proteins against the electrophilic attack.[12] The environment around the carotenoid is extremely important as it can change the carotenoid's structure and reactivity.

Typical carotenoids contain 40 carbon atoms (C40) generated by the condensation of eight C5 isoprenoid units. Most carotenoids in nature are derived from the C40 isoprenoid skeleton which can be modified through cyclization, rearrangement, addition, elimination, and substitution. For example, in higher plants, lycopene (Figure 1.1) is competitively cyclized to form β-carotene with lycopene β-cyclase (LCY-b), or α-carotene with lycopene ε-cyclase (LCY-e) and LCY-b.[13,17] The structure of β-carotene (Figure 1.1) consists of nine conjugated chain double bonds terminated by two cyclic β-ionone rings, where the C5-6 and C5′-6′ double bonds are conjugated to the polyene chain to extend conjugation to eleven conjugated double bonds. The structure of α-carotene (Figure 1.1) consists of nine conjugated chain double bonds terminated at one end with a β-ionone ring, and to the opposite end by an ε ring; this leads to only ten conjugated double bonds in α-carotene. Cyclization of both ends of lycopene with ε-ionone rings is very uncommon in nature. α-carotene and β-carotene are the two structural isomers from which oxygenated carotenoids are derived. For example, β-cryptoxanthin has a similar structure to β-carotene, plus a hydroxyl group on C3, while zeaxanthin has symmetrically two hydroxyl groups at C3 and C3′ positions. In photosynthetic tissues, two sequential hydroxylations of the β rings produce first β-cryptoxanthin, and then zeaxanthin. This reaction is normally catalyzed by carotene β-hydroxylase enzymes of the non-heme di-iron type.[14] Lutein, which is a structural isomer of zeaxanthin, has a structure similar to that of α-carotene, with the two hydroxyl groups positioned symmetrically. An enzyme, CYP97C, can first hydroxylate the ε ring of α-carotene to produce α-cryptoxanthin and then another enzyme CYP97A can hydroxylate the β ring of α-cryptoxanthin to produce lutein.[14] Enzymatic breakage of carotenoids can produce biologically active molecules in both animals and plants. They can also be associated with other biologically active molecules like lipids, proteins or sugars, which change their physical and chemical properties and influence their biological roles. Carotenoids are hydrophobic molecules and are thus located in lipophilic regions of the cells. They can form complexes with proteins, for example, to move through an aqueous environment. Carotenoids complexes with proteins called carotenoproteins are water soluble, and they appear to stabilize carotenoids. The novel type of astaxanthin-binding proteins discovered recently in two green algae strains are proposed to be involved in a unique function of photooxidative stress protection in plants.[15] Carotenoid-binding proteins present in human macula, like zeaxanthin-binding protein and lutein-binding protein, were identified and characterized. The proteins were related to the transport and uptake of macular pigment carotenoids into the human retina.[16] Examples of carotenoid complexes with sugars will be discussed in detail in Chapter 5.

Oxidative cleavage of C40 carotenoids at one or both ends produces smaller molecules such as apocarotenoids, composed of carbon skeletons with fewer than 40 carbons. There are about 120 kinds of apocarotenoids present in some species of plants and animals. Some carotenoids have more than 40 carbons in their structure and are called higher carotenoids. There are about 40 kinds of higher carotenoids with a 45- or 50-carbon skeleton present in some species of archaea.[17] Both carotenoids and their enzymatic cleavage products are associated with other processes positively impacting human health.[14] Because of the increased awareness of the involvement of carotenoids in many biochemical and biological processes, and their association with health benefits, characterization of the oxidation intermediates and their mode of formation and reactions are of considerable importance.

The continuous support from the U.S. Department of Energy from 1986 until May 2014 for carotenoid research at the University of Alabama, Tuscaloosa, enabled properties and structures of carotenoid radicals to be established. Among the numerous physiochemical techniques that have been employed in Prof. Kispert's group since 1986 in the studies of oxidation/reduction reactions of carotenoids are the following: electrochemical measurements such as cyclic voltammetry with DigiSim simulation and

Osteryoung square wave voltammetry, optical spectroscopy (UV, visible, near infrared), optical spectroscopy of carotenoid solutions treated with chemical oxidants (e.g., $FeCl_3$), electron paramagnetic spectroscopy (EPR), simultaneous electrochemical oxidation/reduction and EPR detection, continuous wave and pulsed electron nuclear double resonance (ENDOR), and density functional theory (DFT) calculations. In this book, we will discuss in detail the electrochemical measurements, the EPR measurements in conjunction with DFT calculations, and the implications of these results in photoprotection and formation of carotenoid complexes. While the results of some methods simply confirm those of others, in general, each method addresses different aspects of the redox behavior of carotenoids or the physical properties of the intermediates formed upon the initial electron transfer. Such studies are essential if the complex functions of the ubiquitous natural carotenoids are to be understood. It should be recalled that the survival of photosynthetic plants depends on these compounds, and that carotenoids are also essential nutrients for higher animals to sustain healthy life.

Carotenoids are a diverse class of molecules with unique properties and potential applications in various fields, making them truly extraordinary. Research on carotenoids is ongoing, and scientists continue to investigate the potential benefits of these compounds. A lot more research is needed to understand the complexity of these molecules. It is extremely important to note that only a very small fraction of the about 1,200 carotenoids known to date were studied in depth, or are still currently being studied. There is a great deal of knowledge to be gained from studying and researching the physicochemical properties of carotenoids, their reactions *in vitro* and *in vivo*, and their biological mechanisms of action. In addition, there are infinite practical applications of carotenoids in numerous domains: in human health and nutrition for disease prevention and overall health maintenance, in food and cosmetic industries as natural colorants, in the pharmaceutical industry as dietary supplements and potential treatments for certain illnesses, in aquaculture and poultry feed to improve the color of the resulting products, etc. Our book synthesizes the past 30 years of carotenoid research performed in Prof. Kispert's lab, presents methods used to study carotenoids and carotenoid complexes, and offers a small part of knowledge and expertise in working with these remarkable molecules.

Carotenoids found and isolated from various natural sources participate in electron transfer to the surrounding matrix. For example, as a result of electron transfer in solution, positively charged radical cation ($Car^{\bullet+}$) is generated, and subsequently dication (Car^{2+}), cation ($^{\#}Car^{+}$) and neutral radical ($^{\#}Car^{\bullet}$) intermediates are formed in solution. The oxidation/reduction reactions of carotenoids in solution studied electrochemically, along with detection of their oxidation potentials, are presented in Chapter 2. Chapter 3 will outline the use of DFT molecular orbital calculations to predict the structures of the resulting intermediates that can be formed by various electrochemical processes in solution or detected experimentally on solid supports by EPR, ENDOR and multiple high frequency under high magnetic fields (Chapter 4). Some radical intermediates ($Car^{\bullet+}$ and, hypothetically, $^{\#}Car^{\bullet}$) have been found to provide protection in photosynthesis under varying light conditions (Chapter 5). Important properties of carotenoids in complexes formed with saccharides are presented in Chapter 6. Such supramolecular carotenoid complexes are prepared to overcome some serious shortcomings of carotenoids for applications in pharmaceutical, food and cosmetic industries such as low bioavailability and instability to light, high temperature, oxygen and metal ions. The book concludes with a chapter on suggested resources and instrumentation needed for similar research goals to be met, companies' addresses, web addresses, and a literature summary of isolation methods for new or previously known carotenoids from a wide range of sources.

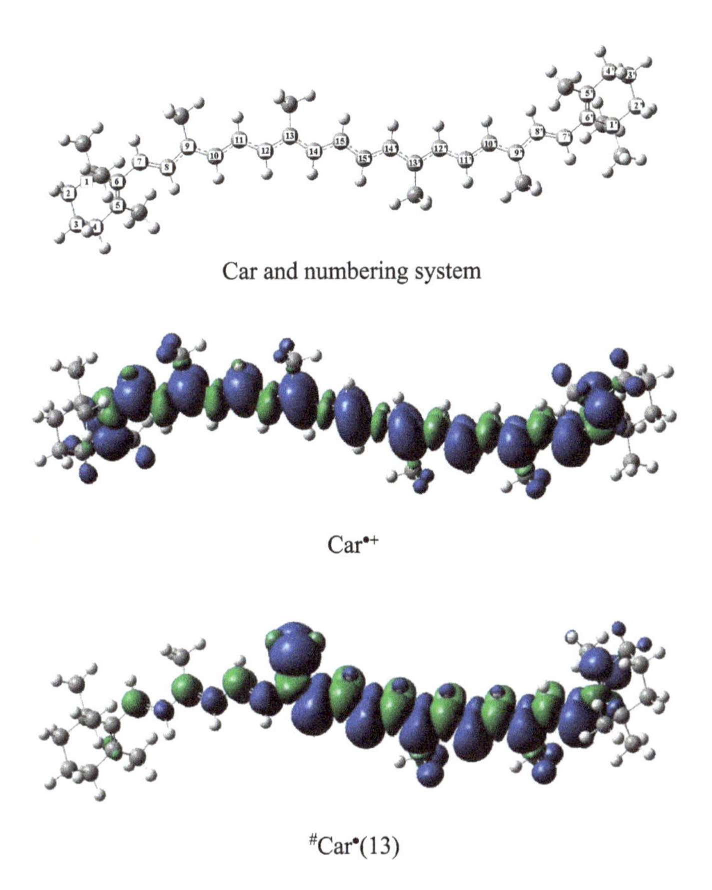

Figure 1.3. The structure of a typical carotenoid with numbering system. The oxidation of the carotenoid molecule Car produces the radical cation according to Car $\rightarrow$ Car$^{\bullet+}$ + e$^-$. The unpaired spin density distribution for the radical cation Car$^{\bullet+}$ determined by DFT is shown. If a proton is lost from the radical cation according to Car$^{\bullet+}$ $\rightarrow$ #Car$^\bullet$ + H$^+$, a neutral radical #Car$^\bullet$ is formed. For example, losing a proton from the methyl group attached at C13 position forms the neutral radical #Car$^\bullet$ (13) (in parenthesis, the position for the proton loss is indicated for different neutral radicals) and the unpaired spin density distribution changes from the position where the proton was lost. Discussed in Chapter 3.

Here is a very brief discussion of the book content with some instructive figures (Figures 1.3–1.8) to inspire the reader to further study the next chapters.

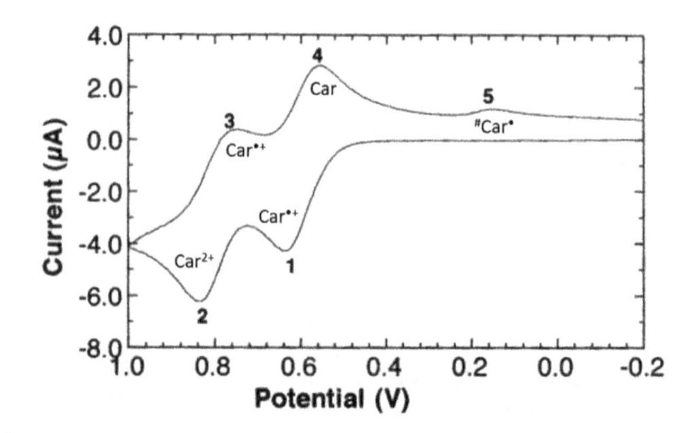

Figure 1.4. A typical cyclic voltammogram for carotenoids. Scanning from right to left during oxidation, the radical cation $Car^{\bullet+}$ (at Peak 1) is formed from the Car molecule by losing an electron. Further oxidation forms the dication, Car^{2+} (at Peak 2). Reducing the dication reforms $Car^{\bullet+}$ (Peak 3) which is further reduced to reform the neutral molecule Car (at Peak 4). Peak 5 is due to formation of the neutral radical, $^{\#}Car^{\bullet}$, as the effect of deprotonation of both radical cation $Car^{\bullet+}$ (weak acid, pK_a approx. 4–7) and dication Car^{2+} (strong acid, pK_a approx. −2). Discussed in Chapter 2.

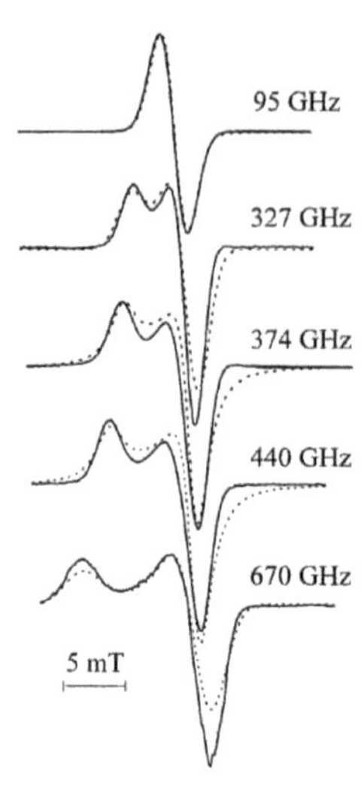

Figure 1.5. The typical EPR spectrum of a carotenoid radical cation $Car^{\bullet+}$ at 9 GHz is composed of just a single Gaussian line with $g_{iso} = 2.0027$, similar to the 95 GHz spectrum in this figure. At high microwave frequency (327 GHz), the symmetrical unresolved EPR line starts to get resolved into two lines indicating the cylindrical symmetry of a π radical cation given by three g components $g_{xx} = g_{yy} = 2.0023$ and $g_{zz} = 2.0032$. Discussed in Chapter 4.

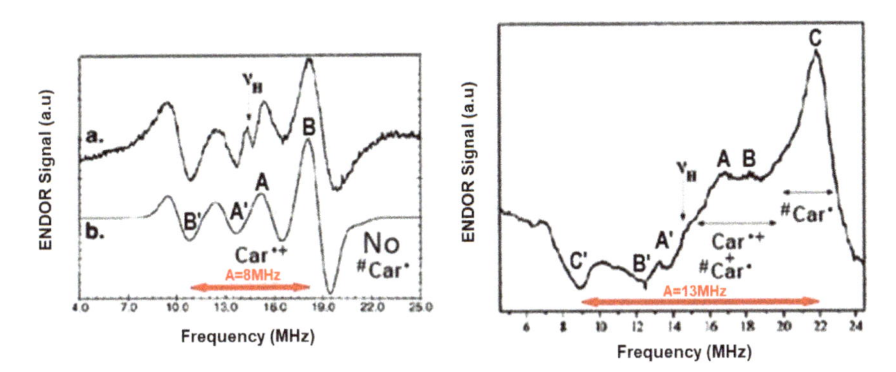

Figure 1.6. More advanced EPR techniques like continuous wave ENDOR can distinguish between Car•+ and #Car• based on their different hyperfine coupling constants. Larger couplings are detected for #Car• (13 MHz) than for Car•+ (8 MHz). The hyperfine coupling A is measured as double the distance around the Larmor frequency υ_H. Any line above approximately 19 MHz in continuous wave ENDOR spectrum is assigned to the neutral radicals #Car•. Up until 2008 the radical chemistry in artificial systems was elucidated and the correlation between quenching ability of carotenoids and neutral radical formation was an inspiration to look for these radical species *in vivo*. Discussed in Chapter 4.

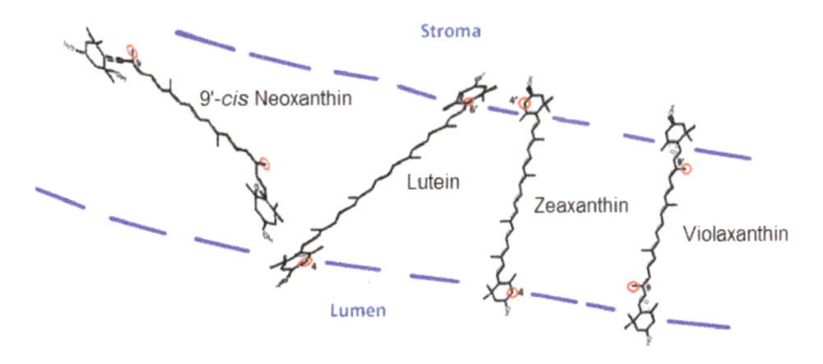

Figure 1.7. Depending on the structure of the carotenoid (positions of proton loss are indicated in red in this picture) and the nature of the environment around it, proton loss from the radical cation Car•+ may or may not generate neutral radicals #Car• that could be essential quenchers of chlorophyll excited states. Discussed in Chapter 6.

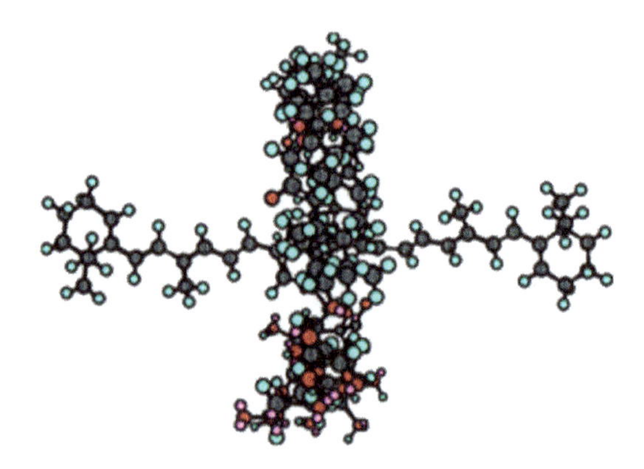

Figure 1.8. Insertion of a carotenoid into other structures forms supramolecular carotenoid complexes with new properties usually resulting in improved stability, solubility, and/or antioxidant activity of the carotenoid. Discussed in Chapter 5.

References

1. Yabuzaki, J. (2017). Carotenoids Database: Structures, chemical fingerprints and distribution among organisms. *Database: the Journal of Biological Databases and Curation, 2017*(1), bax004. https://doi.org/10.1093/database/bax004
2. Wang, D. D., Li, Y., Bhupathiraju, S. N., Rosner, B. A., Sun, Q., Giovannucci, E. L., Rimm, E. B., Manson, J. E., Willett, W. C., Stampfer, M. J., & Hu, F. B. (2021). Fruit and vegetable intake and mortality: Results from 2 prospective cohort studies of US men and women and a meta-analysis of 26 cohort studies. *Circulation, 143*(17), 1642–1654. https://doi.org/10.1161/circulationaha.120.048996
3. World Health Organization. (2003). *Diet, Nutrition and the Prevention of Chronic Diseases: Report of a Joint WHO/FAO Expert Consultation.* WHO Technical Report Series, No. 916. Geneva.
4. Eggersdorfer, M., & Wyss, A. (2018). Carotenoids in human nutrition and health. *Archives of Biochemistry and Biophysics, 652,* 18–26. https://doi.org/10.1016/j.abb.2018.06.001
5. Wilson, L. M., Tharmarajah, S., Jia, Y., Semba, R. D., Schaumberg, D. A., & Robinson, K. A. (2021). The effect of lutein/zeaxanthin intake on human

macular pigment optical density: A systematic review and meta-analysis. *Advances in Nutrition*, *12*(6), 2244–2254. https://doi.org/10.1093/advances/nmab071

6. Obana, A., Gohto, Y., Gellermann, W., Ermakov, I. V., Sasano, H., Seto, T., & Bernstein, P. S. (2019). Skin carotenoid index in a large Japanese population sample. *Scientific Reports*, *9*, 9318. https://doi.org/10.1038/s41598-019-45751-6.
7. Buscemi, S., Corleo, D., Di Pace, F., Petroni, M. L., Satriano, A., & Marchesini, G. (2018). The effect of lutein on eye and extra-eye health. *Nutrients*, *10*(9), 1321. https://doi.org/10.3390/nu10091321
8. Bereczki, I., Papp, H., Kuczmog, A., Madai, M., Nagy, V., Agócs, A., Batta, G., Milánkovits, M., Ostorházi, E., Mitrović, A., Kos, J., Zsigmond, Á., Hajdú, I., Lőrincz, Z., Bajusz, D., Keserű, G. M., Hodek, J., Weber, J., Jakab, F., Herczegh, P., … Borbás, A. (2021). Natural apocarotenoids and their synthetic glycopeptide conjugates inhibit SARS-CoV-2 replication. *Pharmaceuticals*, *14*(11), 1111. https://doi.org/10.3390/ph14111111
9. Meléndez-Martínez, A. J., Mandić, A. I., Bantis, F., Böhm, V., Borge, G., Brnčić, M., Bysted, A., Cano, M. P., Dias, M. G., Elgersma, A., Fikselová, M., García-Alonso, J., Giuffrida, D., Gonçalves, V., Hornero-Méndez, D., Kljak, K., Lavelli, V., Manganaris, G. A., Mapelli-Brahm, P., Marounek, M., … O'Brien, N. (2022). A comprehensive review on carotenoids in foods and feeds: *status quo*, applications, patents, and research needs. *Critical Reviews in Food Science and Nutrition*, *62*(8), 1999–2049. https://doi.org/10.1080/10408398.2020.1867959
10. Lindshield, B. L., & Erdman, J. W. (2010). Carotenoids. In: Milner, J. A., & Romagnolo, D. F. (eds.), *Bioactive Compounds and Cancer: Nutrition and Health*, Humana Press. https://doi.org/10.1007/978-1-60761-627-6_15
11. Nakamura, M., & Sugiura, M. (2019). Health Effects of β-cryptoxanthin and β-cryptoxanthin-enriched Satsuma Mandarin Juice. In: Grumezescu, A. M., & Holban, A. M. (eds.), *Nutrients in Beverages*, Academic Press. https://doi.org/10.1016/B978-0-12-816842-4.00011-3
12. Rice-Evans, C. A., Sampson, J., Bramley, P. M., & Holloway, D. E. (1997). Why do we expect carotenoids to be antioxidants *in vivo*? *Free Radical Research*, *26*(4), 381–398. https://doi.org/10.3109/10715769709097818

13. Misawa, N. (2010). Carotenoids. In: Mander, L., & Lui, H. W. (eds.), *Comprehensive Natural Products II: Chemistry and Biology* Vol. 1, Elsevier.
14. Rodriguez-Concepcion, M., Avalos, J., Bonet, M. L., Boronat, A., Gomez-Gomez, L., Hornero-Mendez, D., Limon, M. C., Meléndez-Martínez, A. J., Olmedilla-Alonso, B., Palou, A., Ribot, J., Rodrigo, M. J., Zacarias, L., & Zhu, C. (2018). A global perspective on carotenoids: Metabolism, biotechnology, and benefits for nutrition and health. *Progress in Lipid Research, 70*, 62–93. https://doi.org/10.1016/j.plipres.2018.04.004
15. Toyoshima, H., Miyata, A., Yoshida, R., Ishige, T., Takaichi, S., & Kawasaki, S. (2021). Distribution of the water-soluble astaxanthin binding Carotenoprotein (AstaP) in *Scenedesmaceae. Marine Drugs, 19*(6), 349. https://doi.org/10.3390/md19060349
16. Li, B., Vachali, P., & Bernstein, P. S. (2010). Human ocular carotenoid-binding proteins. *Photochemical & Photobiological Sciences, 9*(11), 1418–1425. https://doi.org/10.1039/c0pp00126k
17. Maoka, T. (2020). Carotenoids as natural functional pigments. *Journal of Natural Medicines, 74*(1), 1–16. https://doi.org/10.1007/s11418-019-01364-x

2

Electrochemical Studies of Carotenoids in Solution

This chapter summarizes the most representative electrochemical studies[1–24] on carotenoids during the 30 years of research performed in Prof. Kispert's lab at the University of Alabama in Tuscaloosa. These extensive studies of carotenoids have established the electrochemical parameters for numerous natural and synthesized carotenoids, and provided some insight into electron and proton transfer reactions that occur in solution.

The redox potentials of carotenoids in solution can be measured using cyclic voltammetry. A typical cyclic voltammogram (CV) of a carotenoid with its representative peaks and redox potentials that are discussed in this chapter is shown in Figure 2.1. In this example, during the anodic scan, Peak 1 and Peak 2 correspond to the first and second oxidation potentials, E_1^0 and E_2^0, for formation of the radical cation ($Car^{\bullet+}$) and dication (Car^{2+}), respectively, while during the cathodic scan Peak 5 corresponds to the reduction potential E_3^0 for formation of neutral radicals ($^{\#}Car^{\bullet}$).

The origin of all peaks in a CV can be explained based on reactions presented in Scheme 2.1 that take place during an electrochemical oxidation-reduction cycle. In the forward reaction of Eq. (2.1), the carotenoid molecule loses an electron to form the radical cation ($Car^{\bullet+}$) at an oxidation potential E_1^0 (Peak 1). In the forward reaction of Eq. (2.2), the radical cation formed loses a second electron to form the dication (Car^{2+}) at an oxidation potential E_2^0 (Peak 2). The reverse reactions in Eqs. (2.2) and (2.1), or the reduction of the dication and radical cation, respectively, lead to reformation of the radical cation and carotenoid molecule at the position of the two reversible reduction peaks (Peaks 3 and 4, respectively).

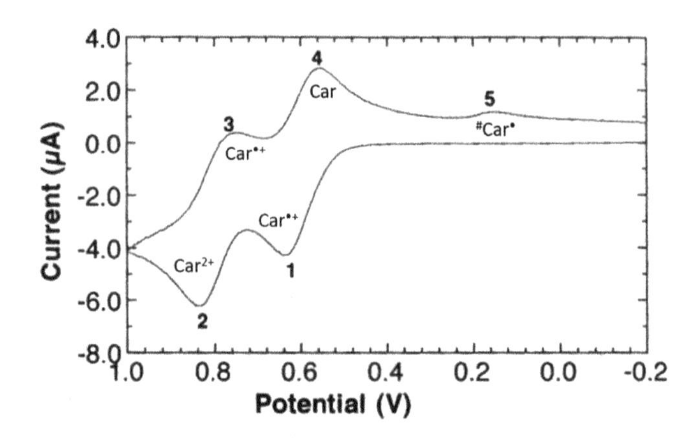

Figure 2.1. Typical CV of a carotenoid.

The third reduction peak, also referred to as the fifth peak (Peak 5) in most electrochemical studies, corresponds to the formation of the carotenoid neutral radical ($^{\#}Car^{\bullet}$) at reduction potential E^0_3. The carotenoid neutral radical, $^{\#}Car^{\bullet}$, is formed by reducing the cation $^{\#}Car^{+}$ (Eq. (2.3), forward reaction) obtained by proton loss from the dication (Eq. (2.5)) which is a strong acid ($pK_a = -2$) and thus can be easily deprotonated. Peak 5 occurs only when dications have been formed, and its intensity depends on the switch potentials that control the amount of dications formed. If the potential is switched before the dications can form, there is no third reduction peak occurring in the CV.[9] The carotenoid neutral radical ($^{\#}Car^{\bullet}$), also known in some studies by the symbol $[Car\text{-}H]^{\bullet}$ to indicate proton loss, can also occur from deprotonation of the radical cation ($Car^{\bullet+}$) (see Eq. (2.6)) because radical cations are weak acids (pK_a = 4–7). Dication Car^{2+} is a strong acid ($pK_a = -2$). Deprotonation of the dication (Eq. (2.5), with deprotonation constant K_{dp}) along with that of the radical cation (Eq. (2.6), with deprotonation constant K'_{dp}) were found to be very important in the mechanism used to simulate the experimental CV to account for the fifth peak. Side reactions Eqs. (2.7)–(2.10) lead to unknown products but their use in simulation was proven to be necessary.[9]

Heterogeneous electrode reactions:

$$\text{Car} \overset{E_1^0}{\rightleftharpoons} \text{Car}^{\bullet+} + \text{e}^- \quad (2.1)$$

$$\text{Car}^{\bullet+} \overset{E_2^0}{\rightleftharpoons} \text{Car}^{2+} + \text{e}^- \quad (2.2)$$

$$^{\#}\text{Car}^{+} + \text{e}^- \overset{E_3^0}{\rightleftharpoons} {^{\#}\text{Car}^{\bullet}} \quad (2.3)$$

Homogeneous reactions:

$$\text{Car}^{2+} + \text{Car} \overset{K_{\text{com}}}{\rightleftharpoons} 2\,\text{Car}^{\bullet+} \quad (2.4)$$

$$\text{Car}^{2+} \overset{K_{\text{dp}}}{\rightleftharpoons} {^{\#}\text{Car}^{+}} + \text{H}^+ \quad (2.5)$$

$$\text{Car}^{\bullet+} \overset{K'_{\text{dp}}}{\rightleftharpoons} {^{\#}\text{Car}^{\bullet}} + \text{H}^+ \quad (2.6)$$

Side reactions:

$$\text{Car}^{\bullet+} \rightarrow \text{Unknown Products} \quad (2.7)$$

$$\text{Car}^{2+} \rightarrow \text{Unknown Products} \quad (2.8)$$

$$^{\#}\text{Car}^{+} \rightarrow \text{Unknown Products} \quad (2.9)$$

$$^{\#}\text{Car}^{\bullet} \rightarrow \text{Unknown Products} \quad (2.10)$$

Scheme 2.1. Reaction mechanisms of carotenoids.

Definition of symbols used in Scheme 2.1:

Car carotenoid molecule

$\text{Car}^{\bullet+}$ carotenoid radical cation

Car^{2+} carotenoid dication

$^{\#}\text{Car}^{+}$ deprotonated carotenoid cation

$^{\#}\text{Car}^{\bullet}$ deprotonated carotenoid radical (also notation $[\text{Car-H}]^{\bullet}$ in some published articles to indicate deprotonation) or so-called neutral radical

The homogeneous reactions (2.4)–(2.6) that take place in solution establish the comproportionation equilibrium constant (K_{com}) between the carotenoid molecule and its radical cation and dication. The value of K_{com} depends on the nature of the substituents on the terminal groups of the carotenoids. With an increase in the electron-withdrawing strength of substituents the separation between the oxidation potentials or the difference $\Delta E = E_1^0 - E_2^0$ increases,[18] which also increases K_{com}.[11] The value of K_{com} varies over 3 orders of magnitude, from ~3.5 for β-carotene[7] which contains cyclohexene electron-donating groups to 1,000 for 7′-apo-7′,7′-dicyano-β-carotene which contains two strongly electron-withdrawing cyano groups. A small K_{com} like in the case of β-carotene that has a low oxidation potential means dications are formed in solution, while larger K_{com} means radical cations are favored in solution, usually occurring for those carotenoids with higher oxidation potential. If a carotenoid is difficult to oxidize (higher oxidation potentials for dication formation), it is easier to reduce the deprotonated dication to form carotenoid neutral radicals.[7] It was also shown[7] that electron-donating and -accepting groups present in a terminal phenyl-substituted carotenoid strongly influence the oxidation potential of these carotenoids and the stability of the cation radicals and dications in dichloromethane solution. The stronger the electron-accepting nature of the substituent, the more stable are the cation radicals and the less stable are the dications in dichloromethane solution.

Reviews of the measured oxidation potentials, K_{com} and other electrochemical parameters for the naturally occurring carotenoids[19] and for some artificially prepared carotenoids[18] as a function of substituents and conjugation length have been published in 1999 and then in 2000 by Liu *et al.*[18,19] The first oxidation potential of carotenoid molecule (E_1^0), which corresponds to the formation of cation radicals, varied from 0.500 V to 0.720 V vs. saturated calomel electrode (SCE). The second oxidation potential of carotenoid cation radical (E_2^0), which corresponds to the formation of a dication, varied from 0.520 V to 0.950 V vs. SCE.[19]

A more recent publication on astaxanthin in 2014 has discussed[22] the care with which electrochemical measurements must be carried out so that oxidation potentials reported from different labs can be compared. Calibrating the electrochemical measurements with ferrocene[21] (0.528 V vs. SCE) is necessary as the conditions of the electrode can vary and carotenoids are known to strongly adsorb on the electrode surface,[12] making each measurement dependent on the electrode surface. To achieve accurate oxidation potentials the CVs were recorded over different orders of magnitude of sweep and fitted as a function of sweep rate with a simulation program.

The most recent review published in 2020 by Gao *et al.*[24] shows an updated list of first oxidation potentials (E_1^0) of new carotenoids measured, along with corrected values for previously measured potentials of selected carotenoids (Table 2.1). The precise measured values for the oxidation potentials of β-carotene and canthaxanthin by Hapiot *et al.*[21] provided a standard reference for carotenoids which was used in the study. The oxidation potentials for β-carotene in methylene chloride were calibrated vs. ferrocene (E_1^0 = 0.528 V vs. SCE) and the electrochemical cell used a Pt wire as counter electrode, SCE as reference electrode and Au (1 mm) disk as working electrode. For other working electrodes like Pt, a similar behavior was observed. Calibration with ferrocene gave the potentials corrected to SCE (see E_1^0 corrected column in Table 2.1). The redox potential reflects the reactivity of a redox couple. A lower value of E_1^0 indicates the carotenoid is more easily oxidized.

In Table 2.1, all the oxidation potentials referenced to SCE are listed from the lowest corrected value 0.593 V for lycopene to 0.952 V for 7,7′-diapo-7,7′-diphenyl-15,15′-didehydro-carotene, a synthetic carotenoid. A first oxidation potential of 0.940 V for carotenoid 9′-*cis*-bixin is the highest oxidation potential reported to date for a natural carotenoid.[23]

Table 2.1. The oxidation potentials were measured in methylene chloride by the method of cyclic voltammetry. The reference electrode used in the measurements was SCE. Calibration with ferrocene gave the potentials corrected to SCE. In the case of the absence of ferrocene calibration, 0.086 V was added to the potentials reported to obtain the corrected value. The subscript sim following a reference indicates simulation of the CV. Structures of the carotenoids are shown in Figure 2.2.

Carotenoid	E^0_1 (V)	E^0_2 (V)	E^0_3 (V)	$E^0_{1\ corrected}$ (V)	References
Lycopene	0.507	0.524	0.051	0.593	4,19_{sim}
Zeaxanthin	0.530	0.550	0.090	0.616	4,19
β-Carotene	0.540	0.545	0.035	0.626	1–3,5,8,9_{sim},10, 11_{sim},12–14, 17,18,19_{sim}
	0.530	0.560	0.045/0035	0.616	6_{sim},7_{sim}
	0.529	0.605		0.614	not published
	0.634			0.634	21
Isozeaxanthin	0.550	0.570	0.080	0.636	5,6_{sim},18,19
7′-Apo-7′,7′-dimethyl-β-carotene	0.568	0.602	0.090/0.160	0.654	9_{sim},11_{sim},18
7,7′-Diapo-7,7′-diphenyl-carotene	0.571	0.648	0.190	0.657	10–12,18
Echinenone	0.590	0.690	0.110	0.676	5,6_{sim},18,19
α-Carotene	0.596	0.623	0.066	0.682	19_{sim}
Rhodoxanthin	0.655	0.900	0.280	0.741	5,6,18,19
8′-Apo-β-caroten-8′-oic acid	0.686	0.846	0.263	0.772	19_{sim}
Canthaxanthin	0.689	0.894	0.264	0.775	1–3,5,7–9_{sim}, 11_{sim},13, 17,18,19_{sim}
	0.705	0.945	0.250	0.791	6_{sim}
	0.678	0.972		0.761	not published
	0.775			0.775	21
Astaxanthin	0.687	0.904	0.250	0.768	22

Table 2.1. (*Continued*)

Carotenoid	E^0_1 (V)	E^0_2 (V)	E^0_3 (V)	$E^0_{1\ corrected}$ (V)	References
8′-Apo-β-caroten-8′-al	0.730	0.890	0.100	0.816	1–3,5,15,16, 17_{sim},18,19
7′-Apo-7′,7′-dicyano-β-carotene	0.739	0.916	0.238	0.825	5,9_{sim},11_{sim},18
15,15′-Didehydro-β-carotene	0.763	0.766	0.240	0.875	9_{sim},10
	0.875	0.822		0.875	21
Fucoxanthin	0.790	0.820	0.240	0.876	19
9′-*cis*-Bixin	0.940	1.140		0.940	23
7,7′-Diapo-7,7′-diphenyl-15,15′-didehydro-carotene	0.866	0.880	0.350	0.952	10,18

The structures of carotenoids listed in Table 2.1 are shown in Figure 2.2. The conjugation length can be measured by the number of double bonds (NDB) in the conjugated chain. It was noted that carotenoid isomers having fewer conjugated double bonds have higher oxidation potentials. For example, when comparing isomers β-carotene (NBD = 11) and α-carotene (NBD = 10), or rhodoxanthin (NBD = 14) with canthaxanthin (NBD = 13), or zeaxanthin (NBD = 11) with lutein (NBD = 10), they have similar structures but just one less double bond in the conjugation chain results in a higher oxidation potential for α-carotene, canthaxanthin and lutein.

In carotenoids containing the same functional group, decreasing the conjugation length increases the oxidation potential, like in aldehydes or esters with varying conjugation lengths. For example, 8′-apo-β-caroten-8′-al (NBD = 10, shown in Figure 2.2) has a higher oxidation potential than 6′-apo-β-caroten-6′-al (NBD = 11, structure not shown here), which

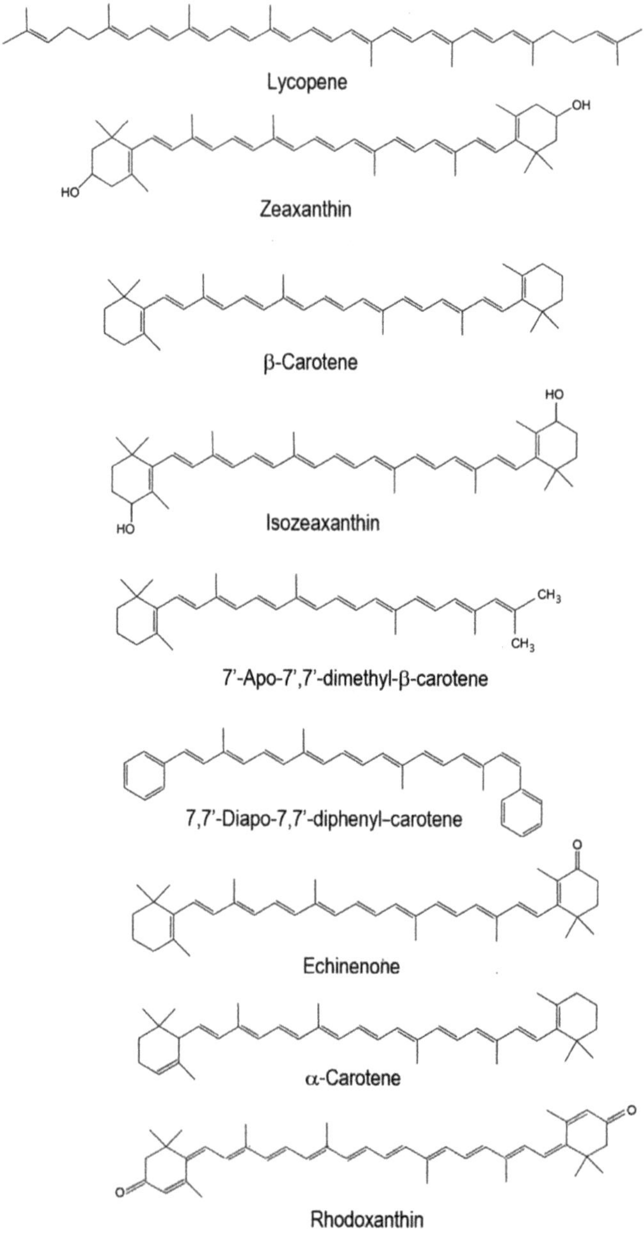

Lycopene

OH

HO

Zeaxanthin

β-Carotene

HO

HO

Isozeaxanthin

CH_3

CH_3

7'-Apo-7',7'-dimethyl-β-carotene

7,7'-Diapo-7,7'-diphenyl-carotene

O

Echinenone

α-Carotene

O

O

Rhodoxanthin

Figure 2.2. Structures of carotenoids from Table 2.1.

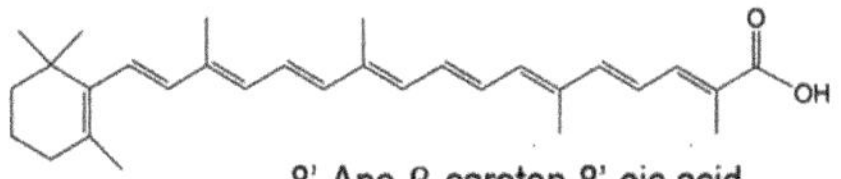

8'-Apo-β-caroten-8'-oic acid

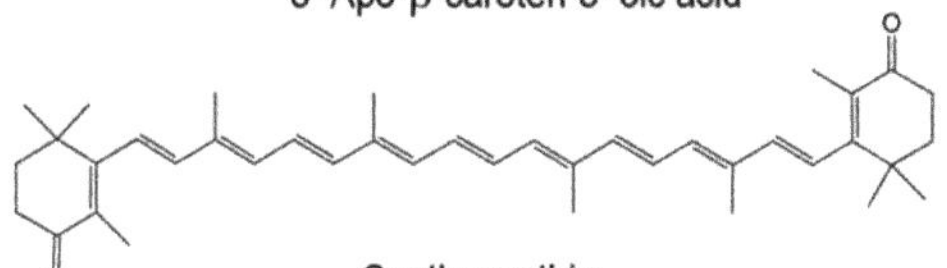

Canthaxanthin

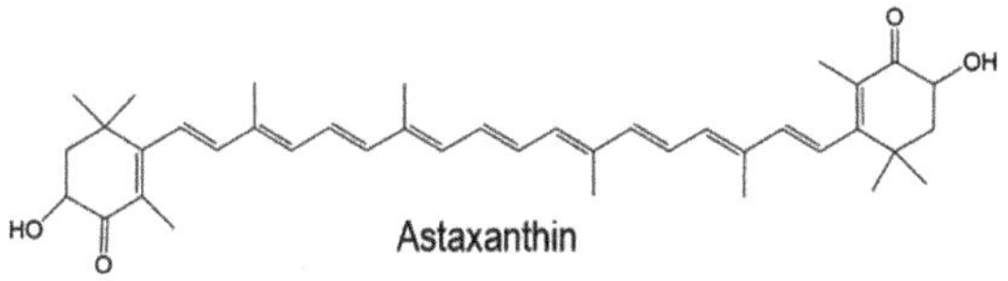

Astaxanthin

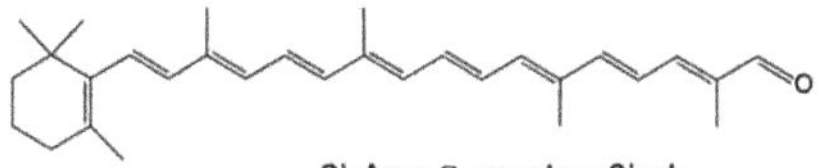

8'-Apo-β-caroten-8'-al

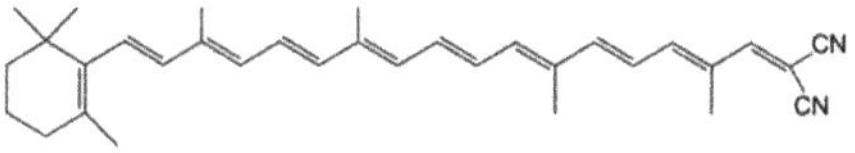

7'-Apo-7',7'-dicyano-β-carotene

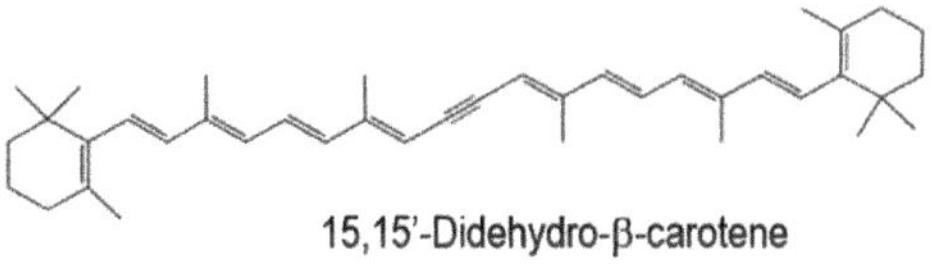

15,15'-Didehydro-β-carotene

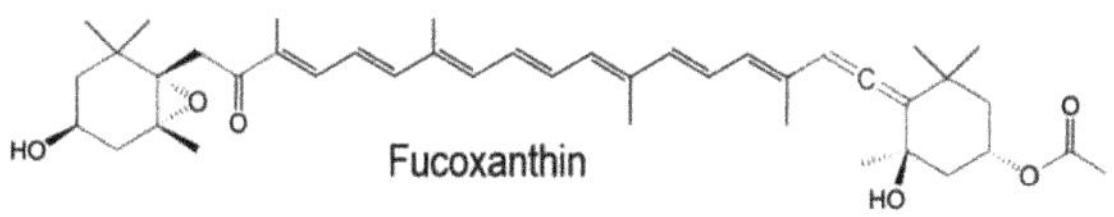

Fucoxanthin

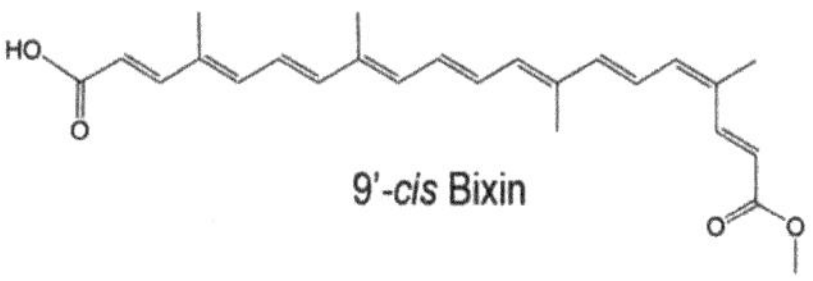

9'-*cis* Bixin

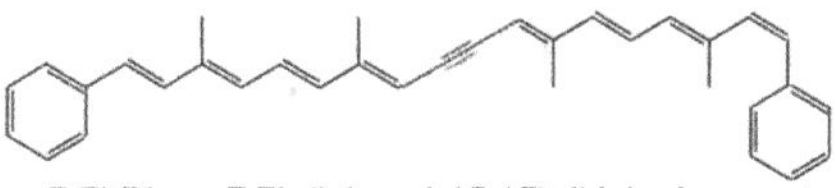

7,7'-Diapo-7,7'-diphenyl-15,15'-didehydro-carotene

Figure 2.2. (*Continued*)

has a higher oxidation potential than 4′-apo-β-caroten-4′-al (NBD = 12, structure not shown here).[18] Synthetic carotenoids containing a triple bond that shortens the conjugation length, like 15,15′-didehydro-β-carotene or 7,7′-diapo-7,7′-diphenyl-15,15′-didehydro-carotene, showed high oxidation potentials.[10] The oxidation potential of a molecule is related to the HOMO (highest occupied molecular orbital) energy of the molecule. The lower the HOMO energy, the higher the oxidation potential. For carotenoids with similar structures, the carotenoid with shorter conjugation length would have lower HOMO energy, and thus higher oxidation potential.

The oxidation potentials are also dependent on the presence of electron-withdrawing groups on carotenoids. The stronger the electron-accepting ability, the greater the proton acidity of the donating proton, which is reflected in a greater oxidation potential. Thus, the more electron-accepting groups a carotenoid contains, the higher the first oxidation potential will be. For example, 9′-*cis* bixin owns the highest first oxidation potential measured up to date (940 mV vs. SCE) for a natural carotenoid (note that 7,7′-diapo-7,7′-diphenyl-15,15′-didehydro-carotene with corrected oxidation potential 0.952 V vs. SCE is synthetic) because it contains four oxygen atoms with strong electronegativity in the conjugated system and a relatively short conjugation chain (NBD = 11). Echinenone contains one less electron-withdrawing carbonyl group than canthaxanthin, and the oxidation potential is much lower than that of canthaxanthin (0.676 vs. 0.775 V), although the conjugation length is slightly shorter than that of canthaxanthin. The higher oxidation potentials for 15,15′-didehydro-β-carotene, fucoxanthin and 7,7′-diapo-7,7′-diphenyl-15,15′-didehydro-carotene (0.875 V, 0.876 V, and 0.952 V, respectively) are attributed to the alkynyl and allenyl groups which are electron-withdrawing groups, in addition to the shorter conjugation lengths.

Synthetic carotenoid 7′-diapo-7,7′-diphenyl-15,15′-didehydro-carotene has two electron-withdrawing phenyl groups plus an alkynyl group to give the highest oxidation potential, 0.952 V vs. SCE. Also, when comparing 15,15′-didehydro-β-carotene with β-carotene with the same conjugation lengths (NBD = 11), it becomes obvious that the presence of the two alkynyl carbons results in much higher oxidation potential than β-carotene (0.875 V vs. 0.634 V).

The first oxidation potential of a carotenoid is reduced if it contains one or more electron-donating groups. For example, when comparing 7′-apo-7′,7′-dimethyl-β-carotene with 7′-apo-7′,7′-dicyano-β-carotene with the same NBD = 10 but different groups, it becomes obvious that 7′-apo-7′,7′-dimethyl-β-carotene which contains two electron-donating methyl groups will have a lower oxidation potential than that of 7′-apo-7′,7′-dicyano-β-carotene which contains two electron-withdrawing cyano groups.

Reversible CVs are ideally needed to obtain accurate redox potentials.[22] Previous published studies have demonstrated the difficulty of obtaining reversible CVs for carotenoids. Electrochemical reversibility and the stability of the cation radicals (and dications) can be estimated from the ratio of the cathodic and anodic currents in CVs. If the ratio is close to 1, the reactions are reversible and the cation radicals (or dications) are relatively stable. Reactions (2.1) or (2.2) of many carotenoid systems are quasi-reversible, that is, the cation radicals and dications have a relatively long lifetime. For other compounds, if the cathodic peaks are small (less than 50% of the anodic current), the cation radicals and dications are less stable and decay faster at ambient temperatures.[18] Some asymmetric carotenoids are irreversible due to the less stable nature of their radical cations and dications. Once the symmetry of the carotenoid is altered, an irreversible CV occurs in which the cathodic peaks are very small indicating that dications and radical cations decay faster at room temperature. For example, asymmetric 8′-apo-β-carotene-8′-al and 7′-apo-7′,7′-dicyano-β-carotene both have the cathodic peak 3 (which corresponds to the reduction of Car^{2+} to $Car^{\bullet+}$) greatly decreased indicating the radical cations are unstable and fewer species of Car^{2+} were generated in peak 2 from radical cations. The asymmetry and increasing the withdrawing effect of substituents causes radical cations and dications to be less stable. When the conjugated length factor is added, carotenoids containing an ester, aldehyde or cyano group display irreversible CVs in which peaks 2, 3, and 4 are absent, showing that the stability of radical cations and dications decreases even more.[18] Symmetric carotenoids like β-carotene, canthaxanthin or astaxanthin display a more reversible behavior, excepting the fifth peak.

One drawback in displaying a reversible behavior is dealing with the multitude of reactions listed in Scheme 2.1 which requires working in a dry atmosphere and in the absence of air.[22] Beside this, other factors like the type of aprotic solvent (for example, comparing CH_2Cl_2 and THF), type of electrolyte and preparation of the working electrode influence this behavior. All carotenoids are soluble and their radical cations have the longest lifetime in methylene chloride which can be easily maintained under anhydrous conditions as compared to other organic solvents. Solvents other than anhydrous methylene chloride need to be rigorously dried by established methods including triple vacuum distillation over metals. For example, when HPLC-grade THF bubbled with nitrogen for 30 minutes was used without further purification, irreversible CVs were obtained. It was also noted that dications of carotenoids are more stable in 0.1 M tetrabutylammonium hexafluorophosphate (TBAHFP) than in the presence of the same concentration of tetrabutylammonium perchlorate or tetrabutylammonium tetrafluoroborate, making TBAHFP the best choice in working with carotenoids.[22]

Most CV measurements in the lab at the University of Alabama were performed using anhydrous methylene chloride and 0.1 M TBAHFP, and a three-electrode setup. However, the fifth reduction peak due to formation of neutral radicals was prominent in previous CV measurements. The fifth reduction peak that distorts the CVs of carotenoids was considered a result of rather annoying side reactions that would complicate simulation and it was tried, without success, to be entirely gotten rid of. Dications which are strong acids ($pK_a \approx -2$) and radical cations which are weak acids (pK_a = 4–7) tend to lose a proton in the presence of water and form cations and neutral radicals, respectively (see Eqs. (2.5) and (2.6) in Scheme 2.1). Cations, upon reduction, also form neutral radicals according to Eq. (2.3). Deprotonation of the radical cation, along with that of the dication, was found to be a very important step in the mechanism used to simulate the experimental CV.[9]

The ideal conditions for generating reversible CVs for carotenoids are given below. In the study of astaxanthin and its ester and monoester[22] the experimental conditions needed to generate reversible CVs without the fifth reduction peak were described. As mentioned above, the fifth reduction peak is due to carotenoid neutral radicals that can form as a result of dications and radical cations in the presence of water, thus working in a dry atmosphere would prevent its appearance. All glassware must be dried overnight in an oven at 120°C to extract the water that adheres to the surface of the glassware. The absence of air is also required as reaction between radical cations and oxygen over the minutes of time needed to do electrochemical measurements generates peroxide radicals, an irreversible reaction that confuses the study. Carrying the entire measurement in a dry box proved to be impractical, but with care and speed it could be done on the benchtop in the first 10–15 minutes of preparing the solution. It is important to have all items ready to do the measurements in the shortest practical time. All glassware along with the electrochemical cell, the working and counter electrodes need to be cleaned and rinsed with acetone, held in an oven overnight, and then transferred to the dry box. The carotenoid solution is prepared in the dry box using anhydrous methylene chloride. Anhydrous methylene chloride is an ideal solvent for electrochemistry of carotenoids because their radical cations have the longest lifetime in it. No plastic syringes can be used to draw methylene chloride, or the solvent may extract contaminants from the plastic. The cell was assembled inside the dry box (excepting the reference calomel electrode) and the carotenoid solution was added to it. Any holes in the cell (for the reference electrode and nitrogen stream) were covered before taking the cell outside the dry box. On the benchtop, a soft stream of nitrogen was passed through the cell, and the reference calomel electrode previously rinsed with acetone and dried was added rapidly to avoid air and moisture exposure. The stream of nitrogen should be soft to prevent agitation and evaporation of the carotenoid solution.[22] Polishing of the working electrode is also extremely important for a successful measurement. The working electrode was polished for 3 minutes, rinsed with acetone and then with methylene chloride. The CVs were run in the first 10–15 minutes after polishing the electrode. If working on the bench top, when removing the working electrode for polishing, the hole in the cell left needs to be covered for the entire duration of the polishing process, maintaining the soft stream of nitrogen passing through the solution. The surface of the working electrode must be carefully prepared by vigorous polishing before measurements. For example, in an extreme case of carotenoid adsorption on electrodes, a reported study showed that a 5,400 MW conducting polymer grew on the end of the electrode, resulting in irreversible CVs with an intense reduction peak and making determination of accurate oxidation potentials impossible.[14] The intense reduction peak was due to adsorption of species involved in generation of dications on the surface of electrodes.

All measurements were carried out at room temperature. Dichloromethane solutions of 0.1 M TBAHFP and 1.0 mM carotenoids were prepared in a dry box under nitrogen atmosphere at room temperature. Great care was taken to avoid exposure of sample solutions to moisture and light. The purity of carotenoids can be checked by thin layer chromatography and ^{1}H-NMR spectroscopy. All carotenoid samples should be stored over Drierite, in dark vials, wrapped with parafilm, or in ampules sealed *in vacuo*, and stored in the freezer at −16°C.

A ferrocene test needs to be carried out to establish a reference point (0.448 V vs. SCE)[22] so that redox potentials in different experiments can be compared. However, the potential of ferrocene/ferrocenium redox couple has not been reported in most of the electrochemical studies of carotenoids.

Different disk working electrodes can be used to study carotenoids: glassy carbon (diameter = 3 mm), gold or platinum (both diameter = 1.6 mm). The disk electrode needs to be polished using larger-sized alumina particles (for example, 1 μm) than smaller-sized ones (for example, 0.05 μm). The auxiliary electrode is a platinum wire, and the reference electrode can be Ag/AgCl electrode or SCE. The distance between electrodes is kept to a minimum and they should be at the same height in solution.

The following CVs (and some Osteryoung square-wave voltammograms (OSWV), Figures 2.3–2.21) were taken for different carotenoids in solution (usually methylene chloride) or simulated using DigiSim program.

Lycopene

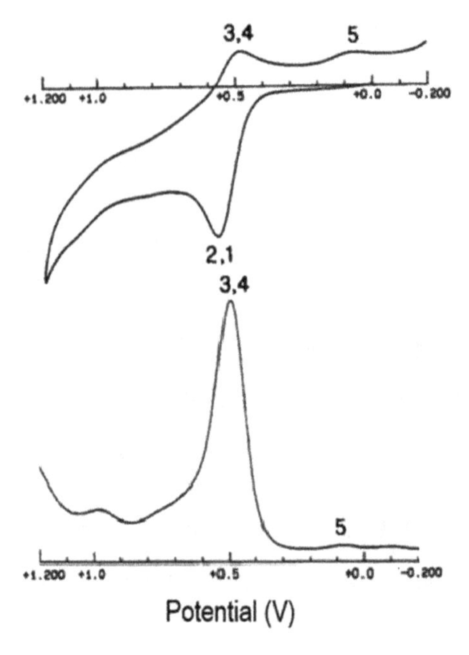

Figure 2.3. CV (top) and OSWV (bottom) of lycopene in methylene chloride with 0.1 M TBAHFP. Scan rate 100 mV/s. Adapted with permission from Figure 6.7 of Reference 4. Simulations were performed by Liu *et al.*, see Table 3 of Reference 19.

Zeaxanthin

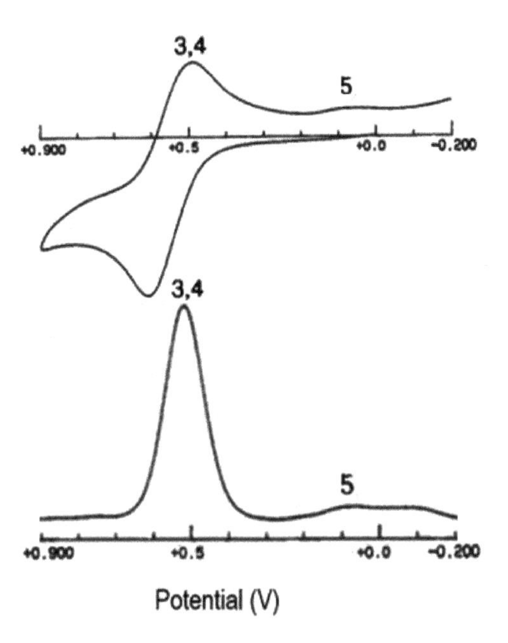

Figure 2.4. CV (top) and OSWV (bottom) of zeaxanthin in methylene chloride with 0.1 M TBAHFP. Scan rate 100 mV/s. Adapted with permission from Figure 6.6 of Reference 4. Values of oxidation potentials were listed by Liu *et al.* in Table 1 of Reference 19.

β-Carotene

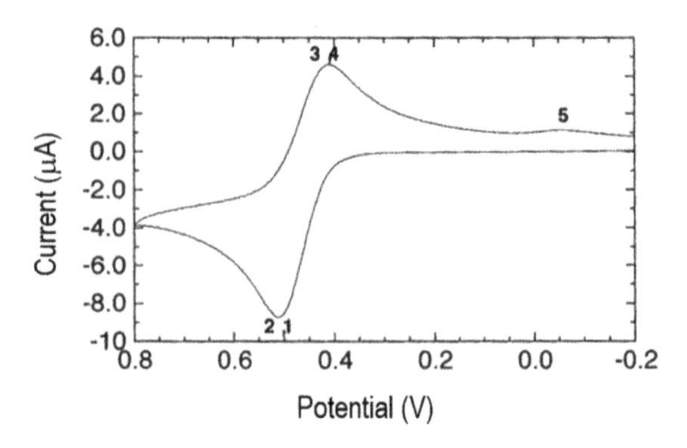

Figure 2.5. CV of β-carotene in methylene chloride and TBAHPF, scan rate 100 mV/s. Adapted with permission from Figure 3.1 of Reference 17. Only one well-defined set of quasi-reversible peaks is observed. This indicates that the electrochemical reaction proceeds in one apparent oxidation step, reactions (2.1) and (2.2) proceed at almost the same time, and the two electrons are transferred in one step. β-carotene undergoes a two-electron transfer reaction. The oxidation potential of neutral β-carotene is only slightly lower than that of the cation radicals; as a result, the two oxidation peaks appear as a single peak. The difference of E_1^0 and E_2^0 was evaluated via simulation by Jeevarajan.[9] For β-carotene, the oxidation potentials of Car (E_1^0) and $Car^{\bullet +}$ (E_2^0) were determined to be 0.540 V and 0.545 V, respectively, while the oxidation potential of $\#Car^{\bullet}$ is 0.035 V (see Table 2.1). Another simulation was performed by Jeevarajan *et al.*[6,7] The oxidation potentials of Car (E_1^0) and $Car^{\bullet +}$ (E_2^0) were determined to be 0.530 V and 0.560 V, respectively,[6,7] while the oxidation potential of $\#Car^{\bullet}$ was 0.045 V (or 0.035 V in Reference 7) vs. SCE.

Izozeaxanthin

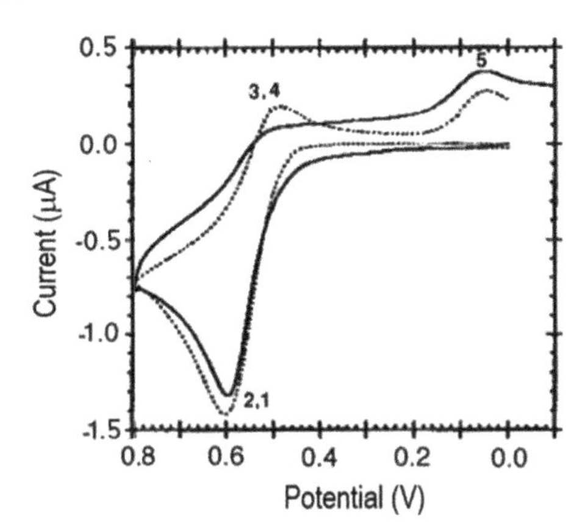

Figure 2.6. Experimental (solid line) and simulated (dashed line) CVs of isozeaxanthin. Adapted with permission from Figure 2 of Reference 6.

7′-Apo-7′,7′-dimethyl-β-carotene

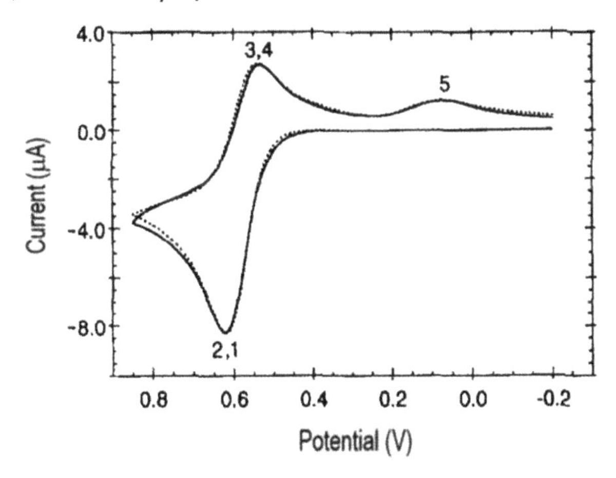

Figure 2.7. Experimental (solid line) and simulated (dashed line) CVs of 7′,7′-dimethyl-7′-β-carotene at 100 mV/s. Reduction potentials E_3^0 were reported for two types of $^{\#}Car^{+}$. Adapted with permission from Figure 3 of Reference 11. Simulation performed by Jeevarajan in Reference 9.

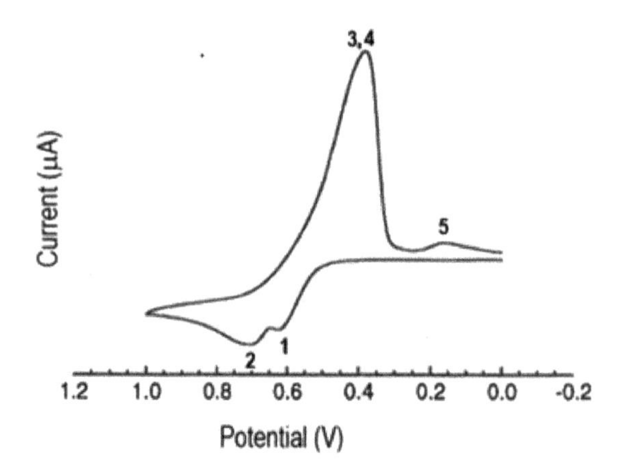

Figure 2.8. CV of 7,7′-diapo-7,7′-diphenyl-carotene in methylene chloride. Adapted with permission from Figure 5 of Reference 10. The anomalous intensity of the cathodic peak 3,4 is due to strong adsorption of the radical cation and dication on the electrode.[12]

Echinenone

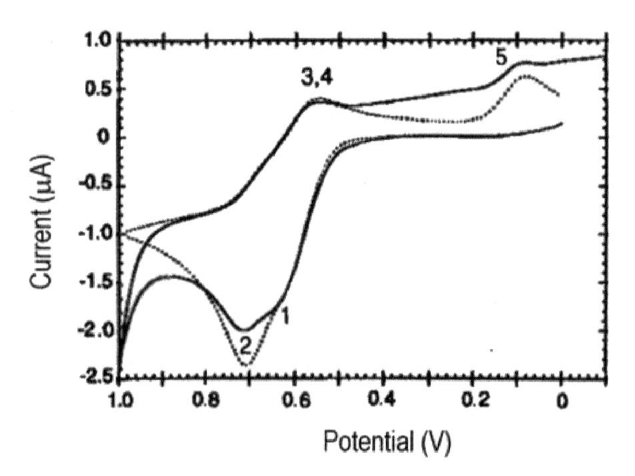

Figure 2.9. Experimental (solid line) and simulated (dashed line) CVs of echinenone. Adapted with permission from Figure 1 of Reference 6.

α-Carotene

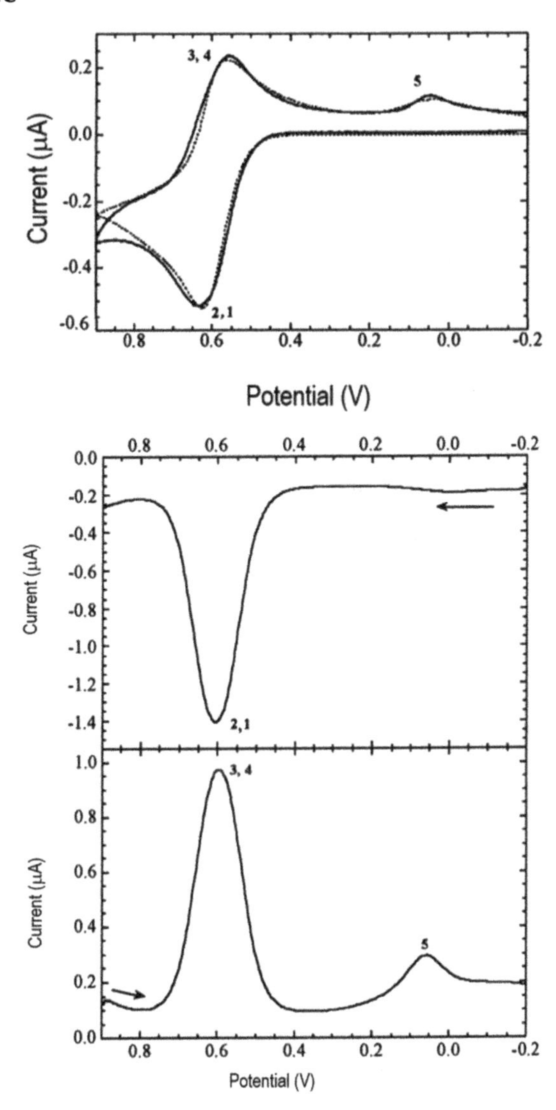

Figure 2.10. Experimental (solid line) and simulated (dotted line) CVs of α-carotene at the scan rate of 0.1 V/s; OSWV in anodic oxidation (top) and cathodic reduction (bottom) process. Pt disk as working electrode in 0.1 mM α-carotene in methylene chloride and 0.1 M TBAHFP. Measurements were conducted in a N_2 dry box. Adapted with permission from Figure 5 of Reference 19.

Rhodoxanthin

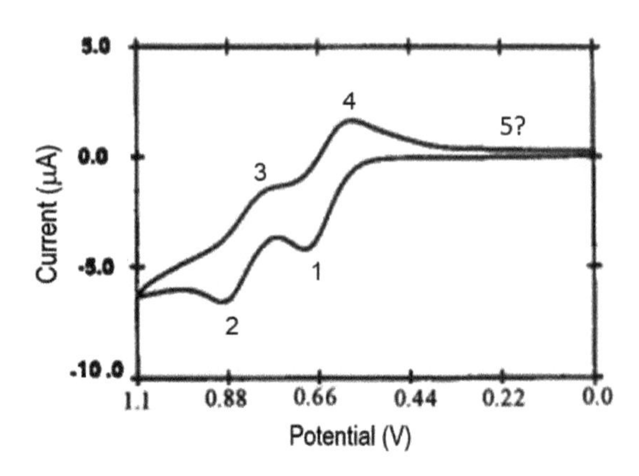

Figure 2.11. CV of rhodoxanthin in methylene chloride. Adapted with permission from Figure 3.8 of Reference 5. Peak 5 is not noticeable.

8′-Apo-β-caroten-8′-oic acid

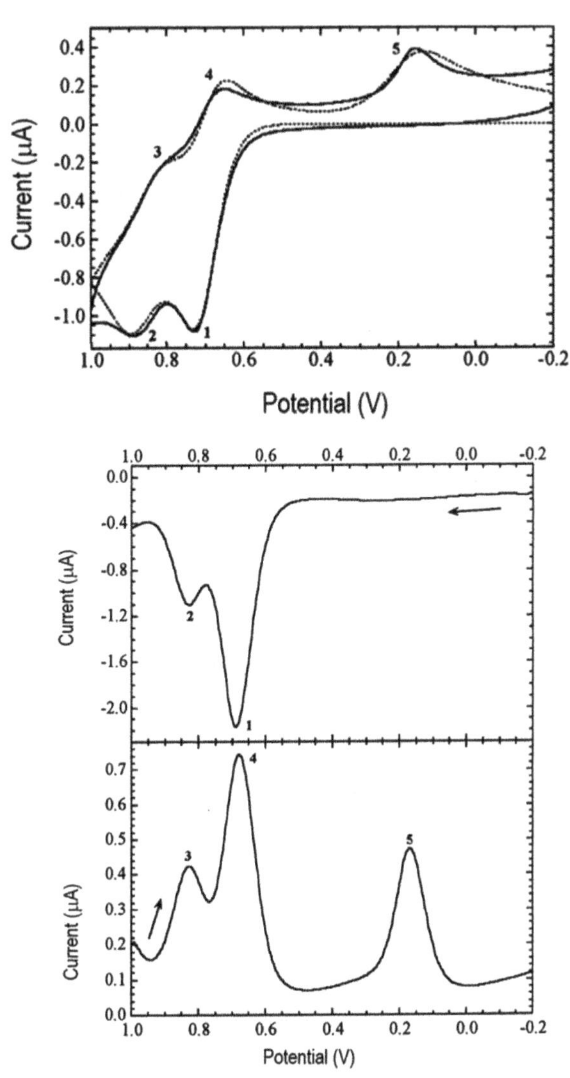

Figure 2.12. Experimental (solid line) and simulated (dotted line) CVs at the scan rate of 0.1 V/s; OSWVs in anodic oxidation (top) and cathodic reduction (bottom) process. Pt disk as working electrode in 0.2 mM 8′-apo-β-caroten-8′-oic acid in methylene chloride and 0.1 M TBAHFP. Measurements were conducted in a N_2 dry box. Adapted with permission from Figure 4 of Reference 19.

Canthaxanthin

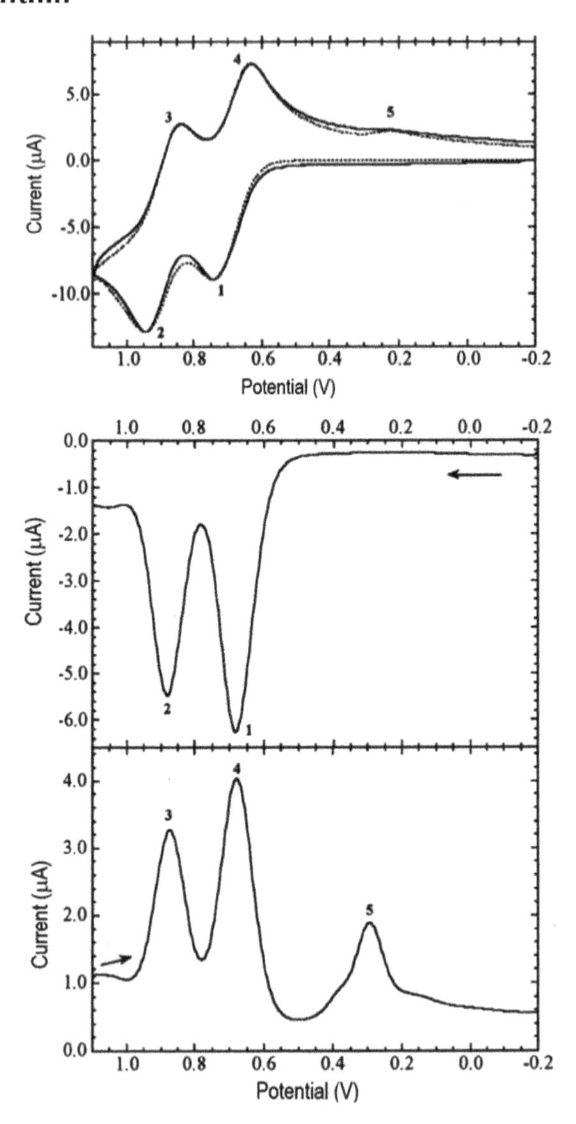

Figure 2.13. Experimental (solid line) and simulated (dotted line) CVs of canthaxanthin at the scan rate of 0.5 V/s; OSWVs in anodic oxidation (top) and cathodic reduction (bottom) process. Pt disk (for CV) and polycrystalline Au disk (for OSWV) as working electrode in 1.0 mM canthaxanthin in methylene chloride and 0.1 M TBAHFP. Adapted with permission from Figure 2 of Reference 19.

Astaxanthin

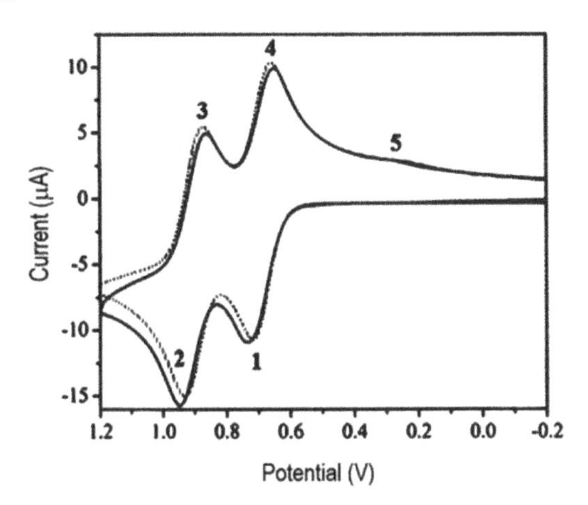

Figure 2.14. Experimental (solid line) CVs of 0.4 mM astaxanthin in methylene chloride and 0.1 M TBAHFP and simulated (dotted line). Adapted with permission from Figure 3 of Reference 22.

8′-Apo-β-caroten-8′-al

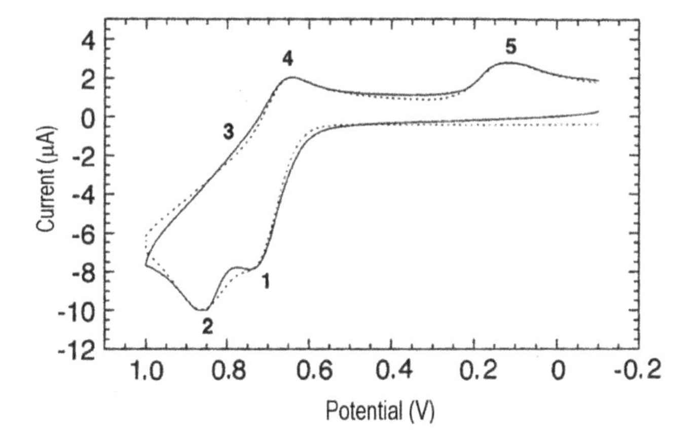

Figure 2.15. Experimental (solid line) and simulated (dotted line) CVs of 8′-apo-β-caroten-8′-al at scan rate 100 mV/s. Adapted with permission from Figure 3.17 of Reference 17.

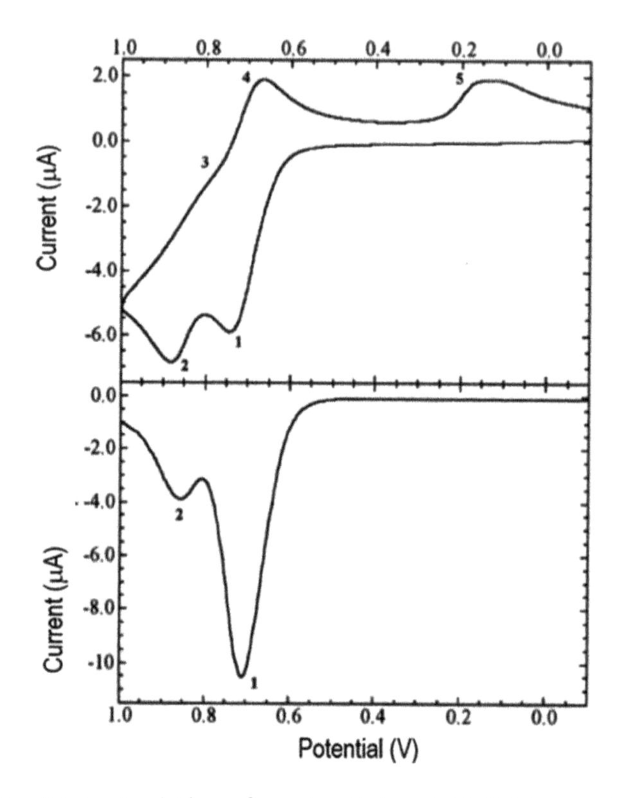

Figure 2.16. CV (top) of 8′-apo-β-caroten-8′-al and OSWV for anodic oxidation (bottom) with Pt disk electrode in 1.0 mM 8′-apo-β-caroten-8′-al in methylene chloride and 0.1 M TBAHFP. Adapted with permission from Figure 3 of Reference 19.

7′-Apo-7′,7′-dicyano-β-carotene

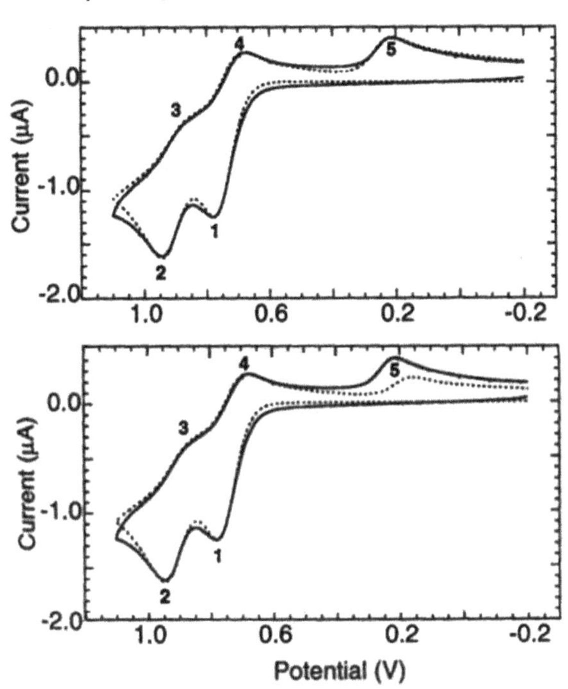

Figure 2.17. Experimental and simulated (dashed line) CV of 7′-apo-7′,7′-dicyano-β-carotene vs. SCE at 100 mV/s vs. SCE. Top CV includes in the simulation the deprotonation of the radical cation. The bottom CV does not include deprotonation of the radical cation and Peak 5 was not satisfactorily reproduced. The deprotonation of the radical cation was found to be a very important step in the mechanism used for simulation of the experimental CV. Adapted with permission from Figure 23 of Reference 9.

15,15′-Didehydro-β-carotene

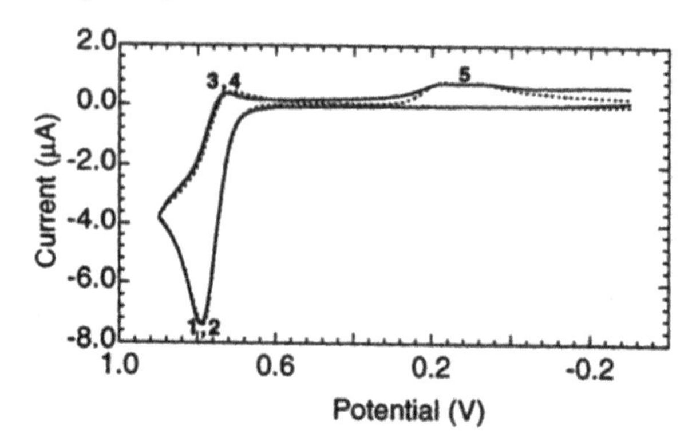

Figure 2.18. CV of 15,15′-didehydro-β-carotene vs. SCE at scan rate of 100 mV/s. Adapted with permission from Figure 4.8 of Reference 9. The broad peak at 5 implies that there is more than one site of deprotonation leading to different cation $^{\#}Car^{+}$ species (see Figure 2.19 for explanation).

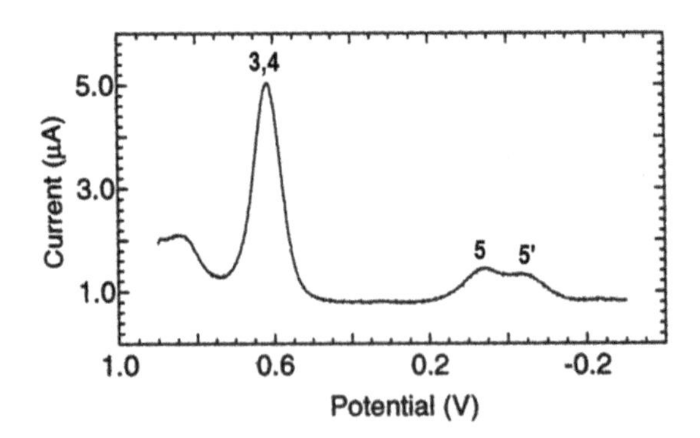

Figure 2.19. OSWV of cathodic reduction of 15,15′-didehydro-β-carotene vs. SCE shows a single peak (3,4) consistent with two electron transfer process in one step. OSWV also shows two partially separated peaks (5,5′) in the low potential region suggesting the presence of more than one reducible species. This implies that there is more than one site of deprotonation leading to different $^{\#}Car^{+}$ species with slightly varying reduction potentials. $^{\#}Car^{+}$ species is formed from the dication and this species undergoes reduction at the potential marked by peak 5. Adapted with permission from Figure 4.2 of Reference 9.

9′-*cis*-Bixin

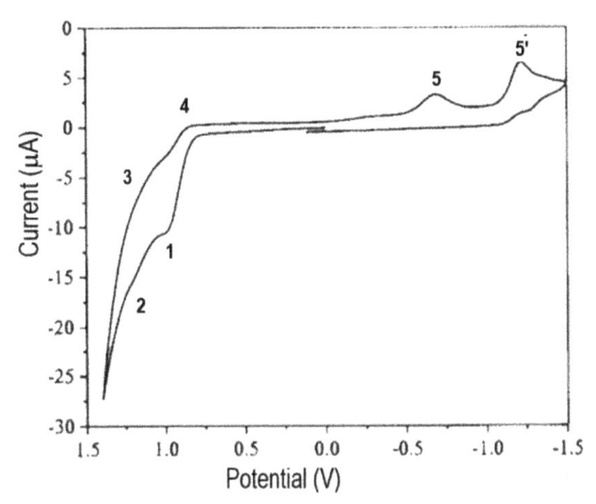

Figure 2.20. CV of 9′-*cis*-bixin at a scan rate of 100 mV/s using THF distilled in the lab and kept on activated 3 Å molecular sieves overnight. The oxidation is a two-electron process with oxidation potentials at ~0.94 and ~1.14 V vs SCE (reference to ferrocene at 0.528 V) in THF. Two reduction potentials were found to occur at ~ −0.69 V and ~ −1.22 V vs SCE. This implies that there is more than one site of deprotonation leading to different $^{\#}Car^{+}$ species with varying reduction potentials. $^{\#}Car^{+}$ species is formed from the dication and this species undergoes reduction at the potential marked by peaks 5 and 6. Adapted with permission from Figure 6 of Reference 23.

7,7′-Diapo-7,7′-diphenyl-15,15′-didehydro-carotene

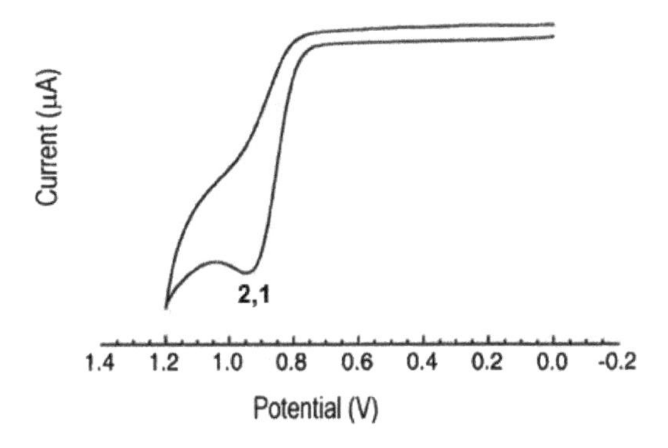

Figure 2.21. CV of 7,7′-diapo-7,7′-diphenyl-15,15′-didehydro-carotene in methylene chloride, scan rate 50 mV/s. Adapted with permission from Figure 6 of Reference 10.

References

1. Grant, J. L., Kramer, V. J., Ding, R., & Kispert, L. D. (1988). Carotenoid cation radicals: electrochemical, optical, and EPR study. *Journal of the American Chemical Society*, *110*, 2151–2157.
2. Khaled, M., Hadjipetrou, A., & Kispert, L. (1990). Electrochemical and electron paramagnetic resonance studies of carotenoid cation radicals and dications: Effect of deuteration. *The Journal of Physical Chemistry*, *94*, 5164–5169.
3. Khaled, M., Hadjipetrou, A., Kispert, L. D., & Allendoerfer, R. D. (1991). Simultaneous electrochemical and electron paramagnetic resonance studies of carotenoid cation radicals and dications. *The Journal of Physical Chemistry*, *95*(6), 2438–2442. https://doi.org/10.1021/j100159a060
4. Chen, X. (1991). An electrochemical, EPR study of carotenoids and high performance liquid chromatography separation of cis-trans isomers of canthaxanthin. Thesis, The University of Alabama, Tuscaloosa, AL.
5. Khaled, M. (1992). Electrochemical, transient EPR, and AM1 excited states studies of carotenoids. PhD dissertation, The University of Alabama, Tuscaloosa, AL.
6. Jeevarajan, A. S., Khaled, M., & Kispert, L. D. (1994). Simultaneous electrochemical and electron paramagnetic resonance studies of keto and hydroxy carotenoids. *Chemical Physics Letters*, *225*, 340–345. https://doi.org/10.1016/0009-2614(94)87091-8
7. Jeevarajan, A. S., Khaled, M., & Kispert, L. D. (1994). Simultaneous electrochemical and electron paramagnetic resonance studies of carotenoids: effect of electron donating and accepting substituents. *The Journal of Physical Chemistry*, *98*(32), 7777–7781. https://doi.org/10.1021/j100083a006
8. Jeevarajan, A. S., Kispert, L. D., & Wu, X. (1994). Spectroelectrochemistry of carotenoids in solution. *Chemical Physics Letters*, *219*, 427–432. https://doi.org/10.1016/0009-2614(94)00121-9
9. Jeevarajan, J. A. (1995). Electrochemical and optical studies of natural and synthetic carotenoids. PhD Dissertation, The University of Alabama, Tuscaloosa, AL.

10. Jeevarajan, J. A., Jeevarajan, A. S., & Kispert, L. D. (1996). Electrochemical, EPR and AM1 studies of acetylenic and ethylenic carotenoids. *Journal of the Chemical Society, Faraday Transactions*, *92*, 1757–1765. https://doi.org/10.1039/FT9969201757
11. Jeevarajan, J. A., & Kispert, L. D. (1996). Electrochemical oxidation of carotenoids containing donor/acceptor substituents. *Journal of Electroanalytical Chemistry*, *411*, 57–66. https://doi.org/10.1016/0022-0728(96)04572-X
12. Gao, G., Jeevarajan, A. S., & Kispert, L. D. (1996). Cyclic voltammetry and spectroelectrochemical studies of cation radical and dication adsorption behavior for 7,7′-diphenyl-7′,7′-diapocarotene. *Journal of Electroanalytical Chemistry*, *411*, 51–56. https://doi.org/10.1016/0022-0728(96)04573-1
13. Gao, G., Deng, Y., & Kispert, L. D. (1997). Photoactivated ferric chloride oxidation of carotenoids by near UV to visible light. *The Journal of Physical Chemistry*, *101*, 7844–7849. https://doi.org/10.1021/jp970630w
14 Gao, G., Wurm, D. B., Kim, Y.-T., & Kispert, L. D. (1997). Electrochemical quartz crystal microbalance, voltammetry, spectroelectrochemical, and microscopic studies of adsorption behavior for (7*E*,7′*Z*)-diphenyl-7,7′-diapocarotene electrochemical oxidation product. *The Journal of Physical Chemistry*, *101*, 2038–2045. https://doi.org/10.1021/jp963741o
15. Wei, C. C., Gao, G., & Kispert, L. D. (1997). Selected cis/trans isomers of carotenoids by bulk electrolysis and iron (III) chloride oxidation. *Journal of the Chemical Society, Perkin Transactions*, *2*, 783–786. https://doi.org/10.1039/A605027A
16. Kispert, L. D., Gao, G., Deng, Y., Konovalov, V., Jeevarajan, A. S., Jeevarajan, J. A., & Hand, E. (1997). Carotenoid radical cations, dications and radical trications. *Acta Chemica Scandinavica*, *51*, 572–578. https://doi.org/10.3891/acta.chem.scand.51-0572
17. Deng, Y. (1999). Carotenoid radical cations and dications studied by electrochemical, optical, and flow injection analysis: Lifetime, extended chain conjugation, and isomerization properties. PhD Dissertation, The University of Alabama, Tuscaloosa, AL.

18. Liu, D., & Kispert, L. D. (1999). Electrochemical aspects of carotenoids. *Recent Research Developments in Electrochemistry, 2*, 139–157.
19. Liu, D., Gao, Y., & Kispert, L. D. (2000). Electrochemical properties of natural carotenoids. *Journal of Electroanalytical Chemistry, 488*(2), 140–150. https://doi.org/10.1016/S0022-0728(00)00205-9
20. Deng, Y., Gao, G., He, Z., & Kispert, L. D. (2000). Effects of polyene chain length and acceptor substituents on the stability of carotenoid radical cations. *The Journal of Physical Chemistry B, 104*, 5651–5656. https://doi.org/10.1021/jp994436g
21. Hapiot, P., Kispert, L. D., Konovalov, V. V., & Savéant, J. M. (2001). Single two-electron transfers vs successive one-electron transfers in polyconjugated systems illustrated by the electrochemical oxidation and reduction of carotenoids. *Journal of the American Chemical Society, 123*(27), 6669–6677. https://doi.org/10.1021/ja0106063
22. Focsan, A. L., Pan, S., & Kispert, L. D. (2014). Electrochemical study of astaxanthin and astaxanthin n-octanoic monoester and diester: tendency to form radicals. *The Journal of Physical Chemistry B, 118*(9), 2331–2339. https://doi.org/10.1021/jp4121436
23. Tay-Agbozo, S., Street, S., & Kispert, L. (2018). The carotenoid bixin found to exhibit the highest measured carotenoid oxidation potential to date consistent with its practical protective use in cosmetics, drugs and food. *Journal of Photochemistry and Photobiology B: Biology, 186*, 1–8. https://doi.org/10.1016/j.jphotobiol.2018.06.016
24. Gao, Y., Focsan, A. L., & Kispert, L. D. (2020). Antioxidant activity in supramolecular carotenoid complexes favored by nonpolar environment and disfavored by hydrogen bonding. *Antioxidants, 9*(7), 625. https://doi.org/10.3390/antiox9070625

3

Density Functional Theory Molecular Orbital Calculations

The combination of molecular orbital calculations such as density functional theory (DFT) calculations with other spectroscopic techniques such as UV-vis, infrared, Raman, NMR or EPR can provide a more comprehensive understanding of carotenoids and their complex interactions. By using DFT, scientists can calculate the electronic structure, geometries, potential energies, and other physicochemical properties of carotenoids, which can be compared with experimental data to validate the results. In this chapter we will discuss the results of DFT studies[1–15] of carotenoid radicals, namely the radical cations ($Car^{\bullet+}$) formed by electron transfer from carotenoid molecules, and the neutral radicals ($^{\#}Car^{\bullet}$) formed by proton loss from the radical cations. These radicals were detected and characterized in solution, powders and *in vivo* in light harvesting complex II. Various EPR techniques such as continuous wave electron nuclear double resonance (CW ENDOR) and pulsed ENDOR methods like Davies and Mims ENDOR in combination with DFT calculations of the hyperfine couplings were used to distinguish and identify these radical species in catalysts containing silanol groups. As a consequence of deprotonation of the carotenoid radical cation, the unpaired electron spin distribution changes so that larger isotropic β-methyl proton couplings occur for the neutral radicals $^{\#}Car^{\bullet}$ (13–16 MHz) than for the radical cation $Car^{\bullet+}$ (7–10 MHz), providing a means to differentiate between these carotenoid radicals. The hypothesis of the photoprotective role of carotenoid neutral radicals formed by proton loss from the radical cation is discussed further in a separate chapter.

Before using DFT calculations in our studies (pre-2006), AM1 and INDO RHF-INDO/SP calculations have proven to be very useful for interpreting the properties of ground-state carotenoid molecules (AM1), dications (AM1) and radical cations (INDO, RHF-INDO/SPDFT) and, surprisingly, even excited states (AM1). The fact that AM1 calculations of excited-state properties (dipole moments, charge density distribution) can be useful to interpret experimental results was demonstrated by agreement with the picosecond results. AM1 calculations have been used as a guide to predict the stability of the isomers, to provide the minimum-energy geometry for the INDO calculation of the unpaired electron density, and to predict the charge density distribution for dications.

With the availability of larger and faster computers, it has been possible to carry out DFT calculations using the Gaussian03 program package[16] on the Cray XD1 computer at the Alabama Supercomputer Center. We have found that small differences in the experimental and INDO/SP calculated hyperfine coupling constants led to errors in the assignment of the radical species in the past (pre-2006).

3.1 DFT Calculations of β-Carotene

In the case of β-carotene (β-car) DFT calculations were carried out[2] with three different functionals (B3LYP, SVWN5 and BPW91) and four different basis sets (3-21G, 6-31G*, DGDZVP2 and 6-31G**). Geometries for the β-carotene radical cation (β-car$^{\bullet+}$) and neutral radicals ($^{\#}$β-car$^{\bullet}$(n) where n represents the position at which proton loss occurs; loss of a proton is indicated by #) were optimized at five levels: B3LYP/3-21G, SVWN5/6-31G*, BPW91/DGDZVP2, B3LYP/6-31G**, and B3LYP/DGDZVP2. Single-point calculations on these geometries were used to generate EPR hyperfine couplings at the B3LYP and B3PW91 levels with the TZP basis set from the Ahlrichs group which has been shown to give good NMR chemical shifts.[17] Filatov and Cremer[18] suggested that improved hyperfine coupling constants can be obtained with the PW91 correlation functional, rather than the LYP correlation functional, because of issues with the amount of spin polarization. However, we did not find any real difference in using the B3PW91 and the B3LYP functionals for the carotenoid radicals,[2] so we used B3LYP for consistency. To model the effect of the environment, the hyperfine coupling constants of the β-methyl protons were also calculated[2] for β-car$^{•+}$ using the self-consistent reaction field (SCRF) approach at the polarizable continuum model level with a dielectric constant of 78.3 for water.[19] DFT calculations on the β-car$^{•+}$ using a SCRF to model a polar water environment showed that the polar environment does not cause significant changes in the proton hyperfine constants from those in the isolated gas-phase molecules.[2] Placing β-carotene radical cation in a polar environment did not cause significant changes in the geometry of the radical cation and thus cannot cause significant changes in the spin distribution.

β-Carotene is a symmetric molecule and calculations for proton loss were performed only for the methyl groups at the unprimed positions C5, C9 and C13 of β-car$^{•+}$, which are equivalent with those for the primed positions C5′, C9′ and C13′ (see Figure 3.1). The electronic energies for the radical cation β-car$^{•+}$, and neutral radicals with proton loss at positions 5, 9 and 13 of the radical cation, namely #β-car$^{•}$(5), #β-car$^{•}$(9) and #β-car$^{•}$(13), calculated at five different levels are given in Table 3.1. Notably, all methods except SVWN5/6-31G* show that the most stable radical structure is that of β-car$^{•+}$ followed by the neutral radicals #β-car$^{•}$(5), #β-car$^{•}$(9) and #β-car$^{•}$(13).

The best geometry as determined by comparison of the hyperfine coupling constants with the experimental couplings from EPR measurements was determined to be B3LYP/6-31G**. The experimental and

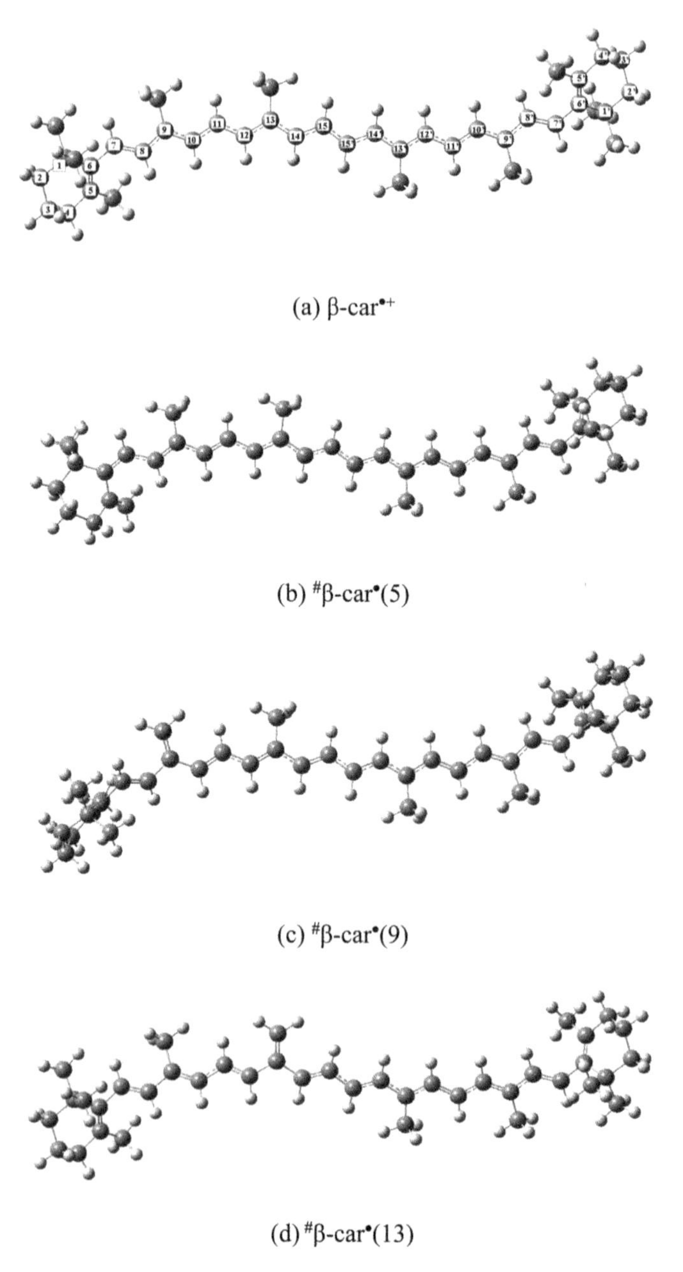

(a) β-car$^{\bullet+}$

(b) $^{\#}$β-car$^{\bullet}$(5)

(c) $^{\#}$β-car$^{\bullet}$(9)

(d) $^{\#}$β-car$^{\bullet}$(13)

Figure 3.1. Side view of the optimized radical structures using B3LYP/6-31G**.

Table 3.1. Total electronic energy (a.u.) of β-carotene radicals with different DFT methods.

DFT Method	β-car$^{\bullet+}$	#β-car$^{\bullet}$(5)	#β-car$^{\bullet}$(9)	#β-car$^{\bullet}$(13)
B3LYP/ 3-21G	−1548.358265	−1547.939350	−1547.930429	−1547.926894
SVWN5/ 6-31G*	−1542.413718	−1541.997079	−1541.960147	−1541.983155
BPW91/ DGDZVP2	−1556.945471	−1556.523060	−1556.511763	−1556.508299
B3LYP/ DGDZVP2	−1557.183923	−1556.767104	−1556.757481	−1556.754409
B3LYP/ 6-31G**	−1556.937349	−1556.513151	−1556.504601	−1556.501584

calculated hyperfine coupling components (in MHz) of carotenoid radical cations from ENDOR measurements available at that time, listed in chronological order, are given in Table 3.2. In 2001, Faller *et al.*[20] showed using ENDOR that the largest hyperfine coupling constant for β-carotene radical cation β-car$^{\bullet+}$ detected in illuminated untreated photosystem II (PS II) or in Mn-depleted PS II at 20 K, and in frozen organic solvent, was not larger than 8.5 MHz (see Table 3.2). The magnitude of the observed couplings was reported as being similar to DFT calculations by Himo,[21] published the same year using B3LYP/6-31G**//B3LYP/3-21G, but very different from our older calculations performed in the early 1990s for the β-carotene radical cation formed by UV photolysis at 77 K on silica gel or Nafion films, which showed large methyl hyperfine couplings on the order of 13–16 MHz.[22,23] The CW ENDOR powder spectrum of β-carotene radical cation formed on silica-alumina was taken later on without UV irradiation: this spectrum lacks the large couplings, even though at that time a 13.0 MHz coupling was assigned to the C13-methyl protons, based on RHF-INDO/SP calculations.[24] The spectral lines labeled A and B in Reference 24 are due only to couplings of 2.0 and 8.5 MHz, with no

Table 3.2. Experimental (from ENDOR measurements) and calculated ^1H hyperfine coupling components (in MHz) for β-carotene radical cation (β-car$^{\bullet+}$).

		Kispert group[22–24]				Faller group[20]		Himo[21]	Kispert group[1]
Position	**Tensor**	**β-car$^{\bullet+}$ (presence of hν) on silica gel[22]**	**RHF-INDO/ SP[22,23]**	**INDO[23]**	**β-car$^{\bullet+}$ (absence of hν) on silica gel[24]**	**β-car$^{\bullet+}$ in PS II[20]**	**β-car$^{\cdot+}$ in organic solvent[20]**	**B3LYP/ 6-31G**// B3LYP/ 6-31G****	**B3LYP/ TZP// B3LYP/ 6-31G****
C5(5′)	A_{11}	2.5			1.9	3.3	2.5	8.4/7.8	5.2
	A_{22}	2.5			1.9	3.3	3.7	6.8/6.5	5.7
	A_{33}	2.8			2.0	3.3	2.5	6.4/6.1	6.8
	$\mathbf{A_{iso}}$	**2.6**	**2.2**	**5.9**	**1.9**	**3.3**	**2.9**	**7.2/6.8**	**5.9**
C9(9′)	A_{11}	8.8			8.2	7.2	6.4	9.2/9.7	6.1
	A_{22}	8.8			8.2	7.2	8.5	8.0/8.3	6.9
	A_{33}	9.2			8.5	7.2	7.0	7.1/7.5	8.1
	$\mathbf{A_{iso}}$	**8.9**	**8.3**	**15.7**	**8.3**	**7.2**	**7.3**	**8.1/8.5**	**7.0**
C13(13′)	A_{11}	12.8/15.9			13.0	5.5	2.6	5.6/5.9	3.8
	A_{22}	12.8/15.9			13.0	5.0	3.8	4.7/5.1	4.6
	A_{33}	14.2/17.0			13.0	4.5	2.6	4.1/4.3	5.4
	$\mathbf{A_{iso}}$	**13.3/16.2**	**12.2/12.5**	**16.3/16.6**	**13.0**	**5.0**	**3.0**	**4.8/5.1**	**4.6**

experimental evidence of very weak spectral lines at frequencies above a 20 MHz ENDOR frequency.

Himo's study[21] used the B3LYP functional and two basis sets, 3-21G and 6-31G**, to calculate the ground state geometries, spin distribution and hyperfine coupling constants of the β-carotene radical cation. The hyperfine coupling constants were obtained by performing single-point calculations on the optimized B3LYP/3-21G and B3LYP/6-31G** geometries with the 6-31G** basis set. They were in the order of 8 MHz, in agreement with the couplings for the carotenoid cation radical in PS II and with that prepared *in vitro* by oxidation with iodine in frozen dichloromethane,[20] and in contradiction to the large coupling constants observed in our measurements of carotenoids adsorbed on silica gel and Nafion films. It was suggested that the 13 MHz coupling could be due to interaction between the carotene radical and the support, and it was shown that the rotation of the head groups can modulate the spin and the hyperfine properties special to the cyclohexene ring.

We thus studied the β-carotene radicals[1,2] using the hybrid DFT method B3LYP[25] and the 6-31G** basis set[26] which proved to be in good agreement with the experimental data. In order to predict the hyperfine couplings, we performed single-point calculations on these geometries at the B3LYP level with the TZP basis set from the Ahlrichs group.[27] The unpaired spin densities were obtained using the AGUI interface from the wavefunctions and spin densities produced by Gaussian 03. The spin density is defined as the difference in the α and β spin densities.

β-Carotene has six methyl groups C5(C5′)-, C9(C9′)- and C13(C13′)-methyl groups in three distinct positions which are similar by symmetry (see structure and numbering system in Figure 3.1). Calculations were carried out on the β-car$^{\bullet+}$ and the neutral carotenoid radicals for the unprimed positions only. Proton loss from each of the methyl groups attached at positions C5, C9 and C13 of the radical cation β-car$^{•+}$ generates neutral radicals #β-car$^{•}$(5), #β-car$^{•}$(9), and #β-car$^{•}$(13), respectively. By symmetry, the hyperfine coupling constants of C5-, C9- and C13-methyl protons of β-car$^{•+}$ are the same as those for C5′-, C9′- and C13′-methyl protons. The optimized structures of the radicals with B3LYP/6-31G** geometry are shown in Figure 3.1.

The geometrical parameters (Table 3.3) and the spin density distribution of β-car$^{•+}$ (Table 3.4) using B3LYP/6-31G** are essentially the same as

Table 3.3. Bond lengths (Å) and angles (°) obtained with B3LYP/6-31G** optimization for β-carotene radical cation and neutral radicals formed by proton loss at the methyl groups.

Bond	β-car$^{•+}$	#β-car$^{•}$(5)	#β-car$^{•}$(9)	#β-car$^{•}$(13)
C1-C2	1.547	1.554	1.546	1.546
C2-C3	1.525	1.531	1.526	1.526
C3-C4	1.525	1.541	1.527	1.527
C4-C5	1.510	1.524	1.515	1.515
C5-C6	1.368	1.482	1.356	1.356
C6-C7	1.455	1.364	1.477	1.476
C7-C8	1.369	1.436	1.347	1.353
C8-C9	1.434	1.375	1.479	1.456
C9-C10	1.390	1.437	1.464	1.365
C10-C11	1.408	1.374	1.364	1.440
C11-C12	1.387	1.419	1.425	1.354
C12-C13	1.413	1.388	1.383	1.472
C13-C14	1.402	1.421	1.426	1.459
C14-C15	1.397	1.387	1.384	1.370
C15-C15′	1.396	1.403	1.406	1.417
C15′-C14′	1.397	1.398	1.395	1.387
C14′-C13′	1.402	1.399	1.402	1.409

Table 3.3. (*Continued*)

Bond	β-car•+	#β-car•(5)	#β-car•(9)	#β-car•(13)
C13′-C12′	1.413	1.421	1.419	1.413
C12′-C11′	1.387	1.379	1.381	1.385
C11′-C10′	1.408	1.422	1.421	1.417
C10′-C9′	1.390	1.377	1.378	1.380
C9′-C8′	1.434	1.447	1.447	1.445
C8′-C7′	1.369	1.358	1.358	1.359
C7′-C6′	1.455	1.473	1.472	1.472
C6′-C5′	1.368	1.358	1.358	1.358
C5′-C4′	1.510	1.514	1.515	1.514
C4′-C3′	1.525	1.527	1.526	1.526
C3′-C2′	1.525	1.526	1.525	1.525
C2′-C1′	1.547	1.546	1.546	1.546
< C5-C6-C7	123.008	123.264	122.382	122.503
< C5-C6-C7-C8	−34.083	−6.481	−48.532	−48.327
< C5′-C6′-C7′	123.008	122.515	122.788	122.700
< C5′-C6′-C7′-C8′	34.083	46.019	−45.424	−45.155

Table 3.4. Unpaired spin density distribution for the β-carotene radical cation and neutral radicals formed by proton loss at the methyl groups.

Molecule	C5	C6	C7	C8	C9	C10	C11	C12	C13	C14	C15
β-car•+	0.12	−0.05	0.17	−0.07	0.16	−0.05	0.13	−0.04	0.09	0.01	0.05
#β-car•(5)	−0.03	0.15	−0.09	0.18	−0.12	0.23	−0.16	0.26	−0.18	0.29	−0.20
#β-car•(9)	−0.01	0.01	−0.03	0.02	−0.09	0.24	−0.15	0.25	−0.18	0.30	−0.20
#β-car•(13)	−0.01	0.01	−0.03	0.02	−0.04	0.03	−0.06	0.06	−0.14	0.33	−0.19
	C15′	**C14′**	**C13′**	**C12′**	**C11′**	**C10′**	**C9′**	**C8′**	**C7′**	**C6′**	**C5′**
β-car•+	0.05	0.01	0.09	−0.04	0.13	−0.05	0.16	−0.07	0.17	−0.05	0.12
#β-car•(5)	0.31	−0.20	0.30	−0.18	0.27	−0.15	0.22	−0.10	0.17	−0.04	0.06
#β-car•(9)	0.32	−0.21	0.32	−0.19	0.28	−0.16	0.23	−0.10	0.18	−0.05	0.06
#β-car•(13)	0.33	−0.21	0.34	−0.21	0.31	−0.17	0.27	−0.12	0.21	−0.05	0.07

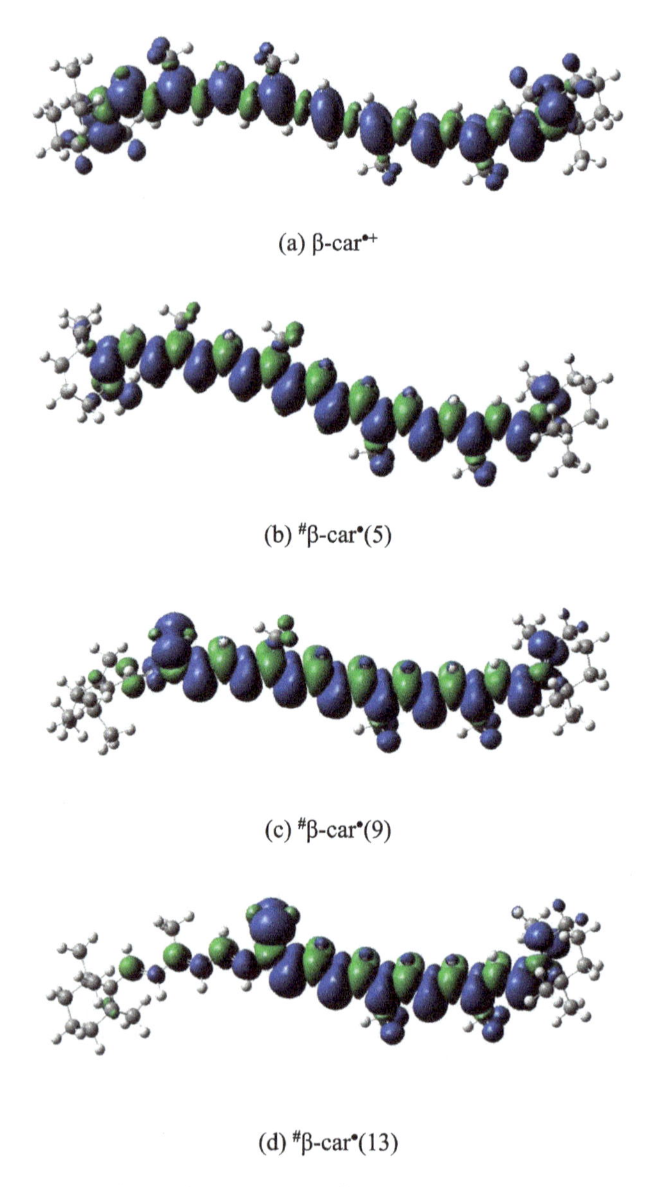

Figure 3.2. Unpaired spin distribution for β-carotene radicals from DFT calculations. Adapted with permission from Figure 4.2 of Reference 2.

those reported previously for the second isoenergetic minima described by Himo with the same method.[21]

The radicals exhibit an odd-alternant spin pattern (Table 3.4), with positive spins alternating with negative spins. However, the pattern from C5 to C15 for the radical cation is opposite to that of the neutral radicals, and the same for all radicals from C15′ to C5′. The spin densities in the C5 to C15 region decrease from $^{\#}\beta$-car$^{\bullet}$(5) to $^{\#}\beta$-car$^{\bullet}$(9) to $^{\#}\beta$-car$^{\bullet}$(13) and increase in the C15′ to C5′ region from β-car$^{\bullet+}$ to $^{\#}\beta$-car$^{\bullet}$(5) to $^{\#}\beta$-car$^{\bullet}$(9) to $^{\#}\beta$-car$^{\bullet}$(13), except that the spin density at C5′ of β-car$^{\bullet+}$ is higher than those of the neutral radicals. The unpaired spin density distribution of the radicals is shown in Figure 3.2. The radical cation has the unpaired spin distributed along the polyene chain from C4 to C4′ (Figure 3.2a). For the neutral radicals the unpaired spin density increases along the polyene chain from the methyl group where the proton was lost (Figure 3.2b, c and d).

The length of delocalization of the spin density for the neutral radicals formed by proton loss at the methyl groups decreases from $^{\#}\beta$-car$^{\bullet}$(5) to $^{\#}\beta$-car$^{\bullet}$(9) to $^{\#}\beta$-car$^{\bullet}$(13), so the relative energies of the neutral radicals are expected to increase in the same direction. Indeed, $^{\#}\beta$-car$^{\bullet}$(5) has the lowest minimum energy among the neutral radicals formed by proton loss at a methyl group and is more stable than $^{\#}\beta$-car$^{\bullet}$(9) by 5.4 kcal/mol, and $^{\#}\beta$-car$^{\bullet}$(9) is more stable than $^{\#}\beta$-car$^{\bullet}$(13) by 1.9 kcal/mol.[2]

3.2 Hyperfine Coupling Constants for β-Carotene

The hyperfine coupling constants of β-carotene can be divided in two classes: isotropic β-methyl proton couplings (A_{iso}) and anisotropic α-proton couplings. The isotropic and anisotropic couplings of β-carotene radical cation and neutral radicals formed by proton loss at the methyl positions calculated with B3LYP/TZP(Ahlrichs)//B3LYP/6-31G** are listed in Table 3.5 and the isotropic methyl couplings for C5(5′), C9(9′)

Table 3.5. DFT-calculated hyperfine coupling constants (MHz) of β-carotene radical cation and neutral radicals using B3LYP/TZP (Ahlrichs)//B3LYP/6-31G**. A_{XX}, A_{YY}, A_{ZZ} — anisotropic coupling tensors for α-protons, A_{iso} — isotropic coupling constants.

Position	β-car•+				#β-car•(5)				#β-car•(9)				#β-car•(13)			
	A_{XX}	A_{YY}	A_{ZZ}	A_{iso}	A_{XX}	A_{YY}	A_{ZZ}	A_{iso}	A_{XX}	A_{YY}	A_{ZZ}	A_{iso}	A_{XX}	A_{YY}	A_{ZZ}	A_{iso}
1				**0.01**				**0.48**				**−0.004**				**−0.01**
				−0.04				**0.17**				**−0.005**				**−0.002**
1′				**0.01**				**0.01**				**0.01**				**0.004**
				0.04				**0.03**				**0.04**				**0.05**
2	−0.13	−0.06	0.44	**0.08**	−0.36	−0.34	0.46	**−0.08**	−0.052	−0.038	0.084	**−0.002**	−0.008	0.001	0.01	**0.001**
	−0.33	−0.25	0.60	**0.01**	−0.34	−0.33	0.25	**−0.14**	−0.06	−0.05	0.07	**−0.01**	−0.04	−0.03	−0.01	**−0.03**
2′	−0.13	−0.06	0.44	**0.08**	−0.20	−0.14	0.37	**0.01**	−0.21	−0.15	0.40	**0.01**	−0.25	−0.16	0.51	**0.03**
	−0.33	−0.25	0.60	**0.01**	0.03	0.09	0.45	**0.19**	0.02	0.09	0.48	**0.20**	0.06	0.14	0.61	**0.27**
3	−0.36	−0.32	0.30	**−0.13**	−0.17	−0.17	0.28	**−0.02**	−0.036	−0.03	0.069	**0.001**	−0.013	−0.005	0.028	**0.003**
	−0.41	−0.31	0.61	**−0.04**	−0.14	−0.10	0.81	**0.19**	−0.06	−0.047	0.11	**0.001**	−0.037	−0.01	0.048	**0.0003**
3′	−0.36	−0.33	0.29	**−0.13**	−0.22	−0.16	0.33	**−0.02**	−0.25	−0.18	0.36	**−0.02**	−0.33	−0.25	0.43	**−0.05**
	−0.41	−0.31	0.61	**−0.04**	−0.18	−0.16	0.19	**−0.05**	−0.20	−0.18	0.21	**−0.06**	−0.28	−0.25	0.23	**−0.10**
4	8.54	8.83	10.21	**9.19**	−1.30	−1.16	−0.51	**−0.99**	−0.39	−0.36	−0.21	**−0.32**	−0.48	−0.42	−0.36	**−0.42**
	11.38	11.61	13.19	**12.06**	−0.29	−0.17	0.43	**−0.01**	−0.29	−0.28	−0.18	**−0.25**	−0.45	−0.40	−0.38	**−0.41**
4′	8.54	8.83	10.21	**9.19**	3.73	3.88	4.57	**4.06**	4.22	4.37	5.15	**4.58**	5.72	5.9	6.91	**6.18**
	11.38	11.61	13.19	**12.06**	3.81	3.84	4.77	**4.14**	4.47	4.51	5.54	**4.84**	6.17	6.22	7.53	**6.64**
5				**5.89**	−2.30	−1.19	0.08	**−1.14**				**−0.17**				**−0.21**
					−1.85	−1.21	−0.57	**−1.21**								

5′				**5.89**				**2.12**				**2.48**				**3.43**
7	−12.04	−8.09	−2.85	**−7.66**	1.25	1.89	3.64	**2.26**	−0.09	0.27	1.99	**0.72**	0.28	0.77	1.38	**0.81**
7′	−12.04	−8.09	−2.85	**−7.66**	−11.14	−7.48	−3.16	**−7.26**	−12.42	−8.38	−3.60	**−8.13**	−17.69	−12.25	−5.88	**−11.94**
8	−0.05	1.07	3.33	**1.45**	−9.72	−7.38	−2.55	**−6.55**	−1.95	−1.63	−0.25	**−1.29**	−0.35	−0.24	−0.22	**−0.27**
8′	−0.05	1.07	3.33	**1.45**	0.84	1.42	4.01	**2.09**	1.04	1.64	4.55	**2.41**	3.03	3.23	7.63	**4.63**
9				**7.05**				**−3.12**	−10.85	−6.28	−1.58	**−6.24**				**−1.14**
									−9.68	−6.47	−2.03	**−6.06**				
9′				**7.05**				**8.44**				**9.23**				**12.03**
10	−0.93	0.02	1.93	**0.34**	−13.20	−9.57	−3.57	**−8.78**	−15.77	−10.58	−3.89	**−10.08**	−0.67	−0.6	−0.33	**−0.53**
10′	−0.93	0.02	1.93	**0.34**	1.87	2.15	6.17	**3.40**	2.20	2.44	6.87	**3.84**	4.65	5.24	11.25	**7.05**
11	−9.66	−7.10	−2.53	**−6.43**	2.18	2.39	6.45	**3.67**	1.70	2.07	6.55	**3.44**	0.19	1.26	4.64	**2.03**
11′	−9.66	−7.10	−2.53	**−6.43**	−17.57	−11.81	−4.97	**−11.45**	−19.15	−12.94	5.53	**−12.54**	−24.99	−17.36	−8.29	**−16.88**
12	−0.8	0.02	2.10	**0.44**	−15.62	−11.44	−4.44	**−10.50**	−15.42	−11.51	−4.36	**−10.43**	−3.90	−3.40	−1.60	**−2.97**
12′	−0.8	0.02	2.10	**0.44**	2.75	2.84	8.15	**4.58**	3.10	3.29	9.02	**5.14**	5.68	6.91	14.04	**8.88**
13				**4.60**				**−5.42**				**−5.32**	−23.51	−14.39	−5.01	**−14.30**
													−21.63	−14.37	−5.34	**−13.78**
13′				**4.60**				**12.48**				**13.31**				**15.87**
14	−2.87	−2.72	0.16	**−1.81**	−19.00	−13.55	−5.22	**−12.59**	−19.69	−14.14	−5.50	**−13.11**	−27.75	−18.95	−7.90	**−18.20**
14′	−2.88	−2.72	0.15	**−1.82**	3.13	3.30	9.14	**5.19**	3.41	3.63	9.77	**5.60**	5.73	6.84	14.25	**8.94**
15	−4.53	−4.02	−0.75	**−3.10**	3.26	3.32	9.05	**5.21**	3.48	3.52	9.44	**5.48**	4.80	5.47	13.52	**7.93**
15’	−4.53	−4.02	−0.75	**−3.10**	−21.22	−14.53	−5.71	**−13.82**	−22.3	−15.38	−6.12	**−14.60**	−26.32	−18.84	−8.21	**−17.79**

and C13(13′) are repeated in the last column in Table 3.2 to show the fit with the experimental hyperfine coupling constants detected in PS II and *in vitro*.[20]

In accordance with the spin density distribution (Table 3.4), the isotropic coupling A_{iso} at C5′ of β-car$^{\bullet+}$ is larger than those of the neutral radicals, and the A_{iso} at C9′ and C13′ of β-car$^{\bullet+}$ are smaller than those of the neutral radicals. The spin densities at C5, C9 and C13 of #β-car$^{\bullet}$(5), #β-car$^{\bullet}$(9), and #β-car$^{\bullet}$(13) are negative, so the proton hyperfine couplings are negative. The CH_2 group protons at C5, C9, and C13 are α-protons which give rise to broadened lines that are usually not detected in powder ENDOR spectra.

The calculated isotropic coupling constants[1] of 5.9, 7.0 and 4.6 for the C5(5′)-, C9(9′)- and C13(13′)-methyl protons for β-car$^{\bullet+}$ that we obtained at the B3LYP/TZP(Ahlrichs)//B3LYP/6-31G** level are slightly smaller than Himo's[21] isotropic coupling constants at the B3LYP/6-31G**// B3LYP/6-31G** level for the two isoenergetic minima of β-car$^{\bullet+}$. However, the isotropic coupling constants for C9(9′)- and C13(13′)-methyl protons obtained at the B3LYP/TZP(Ahlrichs)//B3LYP/6-31G** level fit those detected in PS II and in organic solvent[20] better than those obtained at the B3LYP/6-31G**//B3LYP/6-31G** level. The larger calculated isotropic coupling constant for the C5(5′)-protons than the experimental value was attributed by Himo[21] to rotation of the head groups of the radical cation. The rotation of the head groups can modulate the spin and hyperfine properties, in particular spin delocalization into the cyclohexene ring and the resulting hyperfine coupling constants there. When the dihedral angles <C5-C6-C7-C8 or <C5′-C6′-C7′-C8′ change from 0° to 90°, the isotropic values of C5- or C5′-methyl protons change from 8 to 0 MHz. We can assume that the dihedral angles <C5-C6-C7-C8 and <C5′-C6′-C7′-C8′ of β-car$^{\bullet+}$ in PS II and in organic solvent are slightly different from those calculated for a vacuum because of interactions between β-car$^{\bullet+}$ and the environment. However, the rotation of the head groups in the radical

cation could not account for the 13–16 MHz couplings observed in the solid support environments. These large isotropic couplings were assigned to the neutral radicals of β-carotene, in agreement with their detection in the solid support environments. For example, the calculated 12.48 MHz isotropic methyl proton coupling of #β-car•(5) fits well with the 13.3 MHz obtained in photolyzed β-carotene on silica gel. Another large detected isotropic coupling (16.2 MHz) is in good agreement with the calculated value (15.87 MHz) of the C13′-methyl protons for #β-car•(13).

Based on these observations, we concluded that the β-car•+ species in solid supports can lose protons at different positions to generate neutral radicals. Previous studies[28–31] have shown that in the presence of water, the radical cation easily loses a proton to produce the neutral radical. Since water present in the solid matrix cannot be removed completely during the sample preparation, the neutral radicals can be formed by deprotonation of the radical cations. Deprotonation of the radical cation β-car•+ produced in PS II and dichloromethane[20] is unlikely because there are no water molecules nearby to which the proton can transfer. Our previous study[32] also showed that in a solvent like dichloromethane, two neutral radicals combine to form a didehydrodimer. On solid supports, this reaction does not occur. Diffusion of these neutral radicals is restricted in the solid support environment, and these radicals are relatively stable.

In our previous experiments on silica gel or Nafion film,[22–24] the detection of large isotropic coupling constants up to around 16 MHz suggests production of the neutral radicals by proton loss from the radical cation. Himo[21] suggested that the 13–16 MHz hyperfine coupling constants in the previous ENDOR spectrum may be due to interactions between the radical cation and the solid supports. However, we determined them as being due to the formation of the neutral radicals discussed above, since the isotropic hyperfine coupling constants of β-car•+ in water do not show these large couplings. Our SCRF calculations of β-car•+ in water showed that the optimized structure, the spin distributions and the hyperfine coupling constants of β-car$^{\bullet+}$ in water are very similar to those calculated in vacuum.[2] The calculated hyperfine coupling constants for β-car$^{\bullet+}$ in water, when compared to those in vacuum, are smaller for C5(5′)- and higher for C9(9′)- and C13(13′)-methyl protons, but they are very similar and on the order of 8 MHz. This indicates that the interaction between β-car$^{\bullet+}$ and polar environments will not lead to significant changes in the spin distribution or the associated coupling constants.

3.3 Comparison of Simulated CW ENDOR Spectrum of β-Carotene Radicals to the Experimental Spectrum

In frozen solution β-methyl protons give rise to intense and narrow ENDOR lines, while α-protons give rise to broadened lines which usually escape detection. Using the EasySpin program, in Figure 3.3 we simulated the powder ENDOR spectra of β-carotene radicals using B3LYP/TZP(Ahlrichs)//B3LYP/6-31G** calculated isotropic and anisotropic coupling tensors from Table 3.5. Figure 3.3A, B and C contain the simulations of the powder ENDOR spectra of #β-car$^{\bullet}$(5), #β-car$^{\bullet}$(9) and #β-car$^{\bullet}$(13), respectively. For the simulation of each individual spectrum, we used the calculated full tensor (isotropic and anisotropic) of the corresponding neutral radical listed in Table 3.5. In Figure 3.3D is given the spectrum of the sum of all three individual spectra in a 3:1:1 ratio. Figure 3.3F contains the sum of the neutral radicals (Figure 3.3D) and the radical cation (Figure 3.3E) spectra, and it is compared to the experimental data in Figure 3.3G.

Although the ENDOR simulations do not consider any relaxation effects which are clearly evident in the experimental spectrum (Figure 3.3G), the positions of the simulated lines are very instructive in analyzing the spectrum. Figure 3.3A is the simulated powder ENDOR spectrum of the #β-car$^{\bullet}$(5) neutral radical — the peak marked “a” is due to the 12.48 MHz

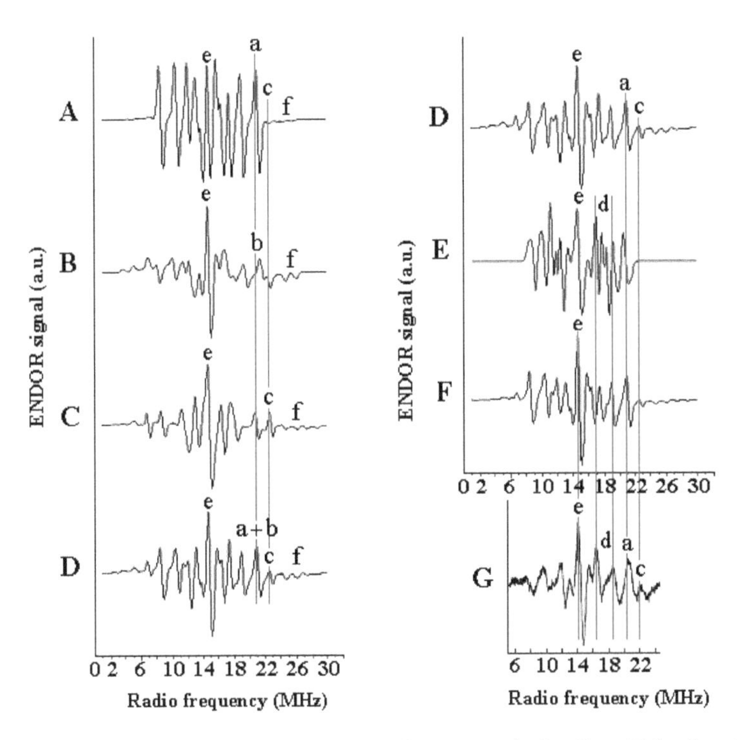

Figure 3.3. Left: Simulated CW ENDOR powder pattern (using linewidth of 0.6 MHz) for (A) #β-car•(5), (B) #β-car•(9) and (C) #β-car•(13) using the DFT hyperfine coupling tensors given in Table 3.5. (D) is the weighted sum of (A), (B), and (C) in the ratio of 3:1:1. (a) is the high frequency line for the 12.48 MHz isotropic methyl coupling for #β-car•(5). (b) is the high frequency line for the 13.3 MHz isotropic methyl coupling for #β-car•(9) and (c) is the 16 MHz isotropic methyl coupling for #β-car•(13). (d) is the line for 8 MHz isotropic methyl coupling. (e) is the location of all couplings under 0.4 MHz and (f) is the location of the high-frequency anisotropic couplings. Right: Simulated ENDOR powder pattern (using linewidth of 0.6 MHz) for (E) β-car•+ and (F) which is the sum of (D) (repeated for reference) and (E). (G) is the relabeled experimental spectrum reported in Figure 4 of Reference 22. Adapted with permission from Figures 4.3 and 4.4 of Reference 2.

C13′-methyl isotropic coupling predicted by DFT. Peak "b" in Figure 3.3B is due to the 13.31 MHz C13′-methyl isotropic coupling for #β-car•(9), and peak "c" in Figure 3.3C is due to the 15.87 MHz isotropic coupling for the C13′-methyl protons of #β-car•(13).

The experimental spectrum, listed in Figure 3.3G, shows the position of these lines a, b and c. It is evident that the experimental spectrum contains spectral lines from all three neutral radicals, although the concentration of $^{\#}\beta$-car$^{\bullet}$(13) appears to be less than that of $^{\#}\beta$-car$^{\bullet}$(5). The simulation of the ENDOR spectrum for β-car$^{\bullet+}$ (Figure 3.3E) shows intense lines (marked "e" in the center of the ENDOR spectrum) due to a large number of small proton couplings less than 1–2 MHz. The intensity of "e" for #β-car$^{\bullet}$(9) provided a way to estimate the concentration of $^{\#}\beta$-car$^{\bullet}$(9) relative to $^{\#}\beta$-car$^{\bullet}$(5), in close agreement with the relative intensity of "e" in Figure 3.3 from Table 3.5. The simulation shown in Figure 3.3F ($^{\#}\beta$-car$^{\bullet}$(5) + $^{\#}\beta$-car$^{\bullet}$(9) + $^{\#}\beta$-car$^{\bullet}$(13)) assumed a 3:1:1 abundance respectively of the three neutral species in this order, and also assumed that the concentration of the neutral radicals is similar to the concentration of β-car$^{\bullet+}$. Comparing Figure 3.3F with the experimental spectrum in Figure 3.3G suggests that the concentration of β-car$^{\bullet+}$ represented by the intensity of line "d" (~8.0 MHz coupling) should be equal to the concentration of $^{\#}\beta$-car$^{\bullet}$(5). It is estimated that upon photolysis, similar amounts of $^{\#}\beta$-car$^{\bullet}$(5), $^{\#}\beta$-car$^{\bullet}$(9) and $^{\#}\beta$-car$^{\bullet}$(13) are produced relative to the amount of β-car$^{\bullet+}$.

The experimental ENDOR spectrum for β-car$^{\bullet+}$ prepared on silica alumina in the absence of UV irradiation (Figure 3 in Reference 24) and the simulation using isotropic couplings of 7.0, 8.3 and 2.6 MHz for methyl proton loss at the C5, C9, and C13 positions of β-car$^{\bullet+}$ are given in Figure 3.4. These values are consistent with the isotropic methyl couplings for β-car$^{\bullet+}$ in PS II and in organic solvent, and with the values from the DFT calculations.

In the absence of light, electron transfer from β-carotene to the matrix occurs and β-car$^{\bullet+}$ is formed.[24] Upon UV irradiation, loss of H^+ from radical cation β-car$^{\bullet+}$ gives rise to the neutral radicals, $^{\#}\beta$-car$^{\bullet}$(n). Our DFT

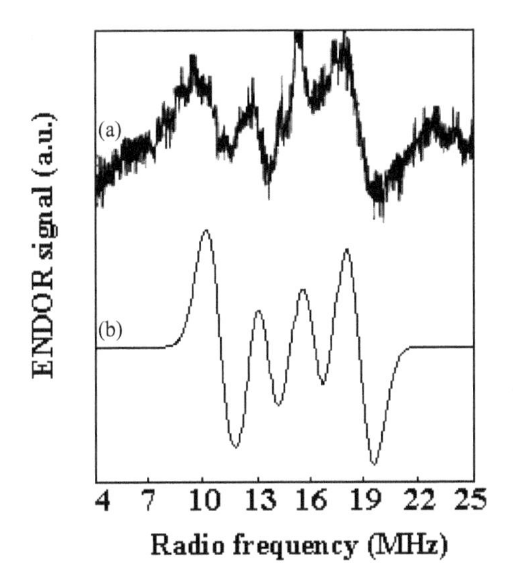

Figure 3.4. (a) Experimental powder ENDOR spectrum (Figure 3 in Reference 24) for the radical cation of β-carotene prepared on a silica alumina matrix in the absence of UV irradiation and (b) simulated powder ENDOR spectrum assuming isotropic proton couplings of 8.3, 7.0 and 2.6 MHz (linewidth 1.5 MHz). Only the rapid averaging of the methyl proton couplings can be detected by powder ENDOR measurements; the ENDOR lines for the anisotropic α-proton couplings are too broad to detect.

calculations (B3LYP/6-31G**) show that the proton affinities (defined as the negative of the enthalpy of the reaction $A + H^+ \rightarrow AH^+$) at 298 K for $^{\#}$β-car$^{\bullet}$(5), $^{\#}$β-car$^{\bullet}$(9) and $^{\#}$β-car$^{\bullet}$(13) are quite high: 267.6, 272.9 and 274.8 kcal/mol, respectively.[2] These gas-phase proton affinities are extremely high for organic compounds, comparable to those for carbenes[33,34] and suggest that the neutral radicals will be strong bases. Another way to consider this reaction is to consider β-car$^{\bullet+}$ as an acid like the imidazolium cation.[35] Of course, the values for loss of a proton from β-car$^{\bullet+}$ correspond to the above basicities as well. Therefore, it is an extremely strong gas-phase acid, much stronger than gas-phase H_2SO_4 and other strong acids.[36]

3.4 Calculations for Proton Loss at C4(4′) Methylene Position of β-Carotene Radical Cation

In addition to proton loss at the methyl groups, loss can occur at C4 or C4′ methylene positions situated on the cyclohexene rings of the radical cation. The neutral radical #β-car•(4 or 4′) formed has the most stable geometry among the other neutral radicals formed by proton loss at the methyl groups, having the unpaired spin density distributed throughout the molecule from C4 to C4′ (see Figure 3.5).

Proton loss from the radical cation at position C4 generates #β-car•(4) neutral radical which is more stable than #β-car•(5) by 4.92 kcal/mol and then #β-car•(9) by 10.29 kcal/mol, followed by #β-car•(13) at a 12.18 kcal/mol energy difference. Table 3.6 shows the electronic energies for these radicals. The radical cation β-car•+ is the most stable, followed by the neutral radicals #β-car•(4), #β-car•(5), #β-car•(9) and #β-car•(13). The hyperfine

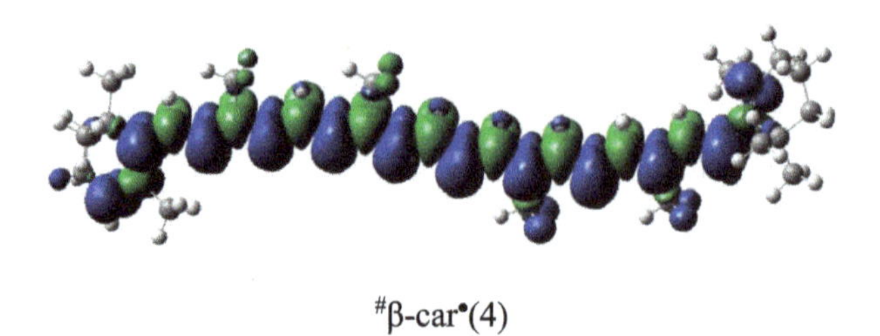

#β-car•(4)

Figure 3.5. Unpaired spin distribution for β-carotene neutral radical formed by proton loss from the radical cation at C4(or 4′) methylene position. Adapted with permission from Reference 2.

Table 3.6. Total electronic energy (a.u.) of β-car•+, #β-car•(4), #β-car•(5), #β-car•(9) and #β-car•(13) with B3LYP/6-31G**.

β-car•+	#β-car•(4)	#β-car•(5)	#β-car•(9)	#β-car•(13)
−1556.937349	−1556.521006	−1556.513151	−1556.504601	−1556.501584

coupling constants of #β-car•(4) (Table 3.7) are on the order of those given by the neutral radicals formed by proton loss at the methyl groups, and similar to those of #β-car•(9) (see Table 3.5). Including these hyperfine coupling constants into the simulation of CW ENDOR spectrum in Figure 3.4 would not change the position of the lines in the spectrum due to the similarity in couplings.

Table 3.7 DFT-calculated isotropic β-methyl proton and anisotropic α-proton hyperfine coupling constants for #β-car•(4) using B3LYP/TZP (Ahlrichs)//B3LYP/6-31G**. A_{XX}, A_{YY}, A_{ZZ} — anisotropic coupling tensors for α-protons, A_{iso} — isotropic coupling constants.

	#β-car•(4)			
Position	A_{XX}	A_{YY}	A_{ZZ}	A_{ISO}
1				**0.43**
				−0.08
1′				**0.05**
				−0.01
2	−0.52	−0.37	0.82	**−0.02**
	0.42	0.59	1.41	**0.81**
2′	−0.20	−0.13	0.40	**0.02**
	0.06	0.13	0.50	**0.23**
3	4.01	4.20	5.26	**4.49**
	8.52	8.70	9.81	**9.01**
3′	−0.26	−0.19	0.33	**−0.04**
	−0.22	−0.20	0.12	**−0.08**
4	−7.35	−4.67	−1.62	**−4.55**
4′	4.45	4.60	5.38	**4.81**
	4.65	4.68	5.71	**5.01**
5				**−1.75**

(*Continued*)

Table 3.7. (*Continued*)

	$^{\#}\beta$-car$^{\bullet}$(4)			
Position	A_{XX}	A_{YY}	A_{ZZ}	A_{ISO}
5′				**2.59**
7	3.23	3.34	6.94	**4.50**
7′	−14.45	−10.04	−4.89	**−9.79**
8	−14.24	−10.46	−4.70	**−9.80**
8′	2.73	2.81	6.43	**3.99**
9				**−5.09**
9′				**9.92**
10	−17.71	−13.00	−5.91	**−12.21**
10′	4.22	5.00	9.97	**6.38**
11	4.92	5.99	11.45	**7.45**
11′	−21.32	−14.82	−7.30	**−14.48**
12	−20.02	−14.85	−6.98	**−13.95**
12′	5.40	6.89	12.98	**8.42**
13				**−7.75**
13′				**13.96**
14	−25.30	−18.98	−9.96	**−18.08**
14′	6.10	7.87	14.50	**9.49**
15	6.34	8.00	14.72	**9.69**
15’	−25.30	−17.86	−8.45	**−17.20**

Deprotonation of the β-carotene radical cation at the C4 methylene position or at the C5-, C9- and C13-methyl groups produces the neutral radicals $^{\#}\beta$-car$^{\bullet}$(4), $^{\#}\beta$-car$^{\bullet}$(5), $^{\#}\beta$-car$^{\bullet}$(9) and $^{\#}\beta$-car$^{\bullet}$(13), respectively. The DFT calculations show that $^{\#}\beta$-car$^{\bullet}$(4) is the most stable radical among the neutral radicals (Table 3.6) due to retention of the π-conjugated system. $^{\#}\beta$-car$^{\bullet}$(5) also retains the π-conjugated system but it has higher energy. #β-car$^{\bullet}$(9) and #β-car$^{\bullet}$(13) are less stable than $^{\#}\beta$-car$^{\bullet}$(5) because the π-conjugated chain is distorted, and the spin is delocalized over only part of the conjugated system (Figure 3.2).

3.5 DFT Calculations for Other Carotenoid Radicals

As described above, DFT calculations[1,2] were first used in our group around 2006 and after to correct previously missed assignments for the carotenoid radical cation by the only available RHF-INDO/SP calculations (pre-2006). Geometries for other carotenoids were optimized with the B3LYP exchange-correlation functional and the 6-31G** basis set,[25] which we have shown to be reasonable for predicting the geometry of β-carotene radicals. Single-point calculations on these geometries were used to predict hyperfine couplings at the B3LYP level with the TZP basis set from the Ahlrichs group,[27] which has been shown to give good EPR parameters for carotenoids. Using DFT-generated couplings as fit to experimental couplings, it was confirmed that the isotropic β-methyl proton hyperfine couplings do not exceed 8–10 MHz for the carotenoid radical cations. We have found some exceptions for asymmetric carotenoids like 8′-apo-β-caroten-8′-al or 7′-apo-7,7′-dicyano-β-carotene or 9′-*cis*-neoxanthin, whose isotropic β-methyl coupling constants could be up to 12 MHz; however, the much larger couplings are due to carotenoid neutral radicals formed by proton loss of the most acidic proton from the radical cation. The largest coupling (around 16 MHz) is the C13′ isotropic coupling of the neutral radical $^{\#}Car^{\bullet}(13)$ (see Table 3.8). DFT calculations explained the large isotropic couplings previously observed in ENDOR measurements of irradiated carotenoids supported on silica gel, Nafion films, silica alumina matrices or incorporated in molecular sieves.

We have concluded that larger couplings of 13–16 MHz determined by DFT and present in ENDOR spectra of photo-irradiated carotenoids on solid supports such as silica gel, Nafion films, silica alumina or incorporated into MCM-41 molecular sieves are attributed to the formation of the neutral radicals. DFT calculations of the minimum energy and structure for proton loss of various carotenoid radicals and EPR detection of the carotenoid neutral radicals stabilized on various solid matrices are summarized in Table 3.9.

Table 3.8. The largest isotropic coupling constants from DFT calculations (B3LYP/TZP (Ahlrichs)//B3LYP/6-31G**) for the radical cation $Car^{\bullet+}$ and neutral radical $^{\#}Car^{\bullet}(13)$ which gives the largest coupling among the neutral radicals.

Carotenoid	Largest isotropic coupling for radical cation $Car^{\bullet+}$	Largest isotropic coupling for neutral radical $^{\#}Car^{\bullet}(13)$	Reference
Astaxanthin	9.17(C9)/9.23(C9′)	15.93(C13′)	9
Zeaxanthin	8.33(C9)/8.26(C9′)	16.10(C13′)	3
β-Carotene	7.05(C9)/7.05(C9′)	15.87(C13′)	1,2
Lycopene	7.84(C5)/7.95(C5′)	15.57(C13′)	2,8
8′-Apo-β-caroten-8′-al	10.46(C9′)/10.74(C13′)	16.70(C13′)	2
7′-Apo-7,7′-dicyano-β-carotene	10.95(C13′)/11.44(C9′)	15.05(C13′)	2
Canthaxanthin	9.36(C9)/9.34(C9′)	16.04(C13′)	2,8
Lutein	8.04(C9)/8.94(C9′)	16.35(C13′)	2,4
Violaxanthin	9.93(C9)/10.00(C9′)	16.38(C13′)	3
9′-*cis*-Neoxanthin	10.41(C9)/12.11(C9′)	16.76(C13′)	5

Table 3.9. Carotenoid neutral radicals isolated on various matrices detected by EPR and characterized by DFT.

Carotenoid	Matrix	Reference
Astaxanthin	MCM-41, Ti-MCM-41, silica alumina	7,9
Zeaxanthin	Silica alumina	3
β-Carotene	Silica alumina, Cu-MCM-41	1,6
Lycopene	Cu-MCM-41, silica alumina	4,8
7′-Apo-7,7′-dicyano- β-carotene	Cu-MCM-41	4
Lutein	Cu-MCM-41, Cu-SBA-15	4
Violaxanthin	Silica alumina	3
9′-*cis*-Neoxanthin	MCM-41	5
Linear carotenoids: Neurosporene Spirilloxanthin	DFT simulations only	8

EPR and DFT studies have identified the radicals formed on these matrices for numerous carotenoids including β-carotene, zeaxanthin, violaxanthin, lutein, neoxanthin and astaxanthin. Upon light illumination, the matrix surfaces are activated, forming proton acceptors. Also, electrochemical studies present radical cations as being weak acids, enabling proton donation and thus forming the carotenoid neutral radical. In such assemblies like the solid matrices described (and possibly in solution), proton loss detected by advanced EPR occurs from different positions of the radical cation.

The neutral radicals ($^{\#}Car^{\bullet}(n)$) can be produced as a result of proton loss from the methyl groups and from the methylene positions of the radical cation, $Car^{\bullet +}$. The most favorable positions for proton loss from the radical cation can be predicted by DFT by calculations of relative energies for the neutral radicals (Table 3.10). In general, the most energetically favorable proton loss from the radical cations occurs from positions on the terminal rings that extends the conjugated system (Figure 3.6). As a consequence of deprotonation, the unpaired electron spin distribution changes so that larger methyl proton hyperfine constants occur for the neutral radicals (13–16 MHz) than for the radical cation as described above.

Several papers[1,3–5,8,9] have been published on the unpaired distribution that occurs for various carotenoids as a function of substituent, symmetry and function. According to Table 3.10, the most preferred proton loss from astaxanthin radical cation is from position C3 or by symmetry from the C3′ on the terminal rings (see Figure 3.6) of the radical cation, which extends conjugation. Deprotonation at C3, and by symmetry at C3′, resulted in the lowest energy neutral radical due to resonance stabilization upon hydroxyl proton migration to the carbonyl group (indicated by the minimum energy $\Delta E = 0$ kcal/mol and delocalization length 26).[9] DFT calculations for astaxanthin determined that proton loss at the terminal rings of the radical cation was favored (including loss of the C3 proton

Table 3.10. Relative energies $\Delta E(n)$ in kcal/mol (calculated relative to the energy minimum) of carotenoid neutral radicals formed by proton loss from the radical cation. Letter "n" indicates the position from which the proton was lost. For example, $\Delta E(4)$ is the relative energy for $Car^{\bullet}(4)$, the neutral radical formed by proton loss from the C4 position and $\Delta E(5)$ is the relative energy for $Car^{\bullet}(5)$, the neutral radical formed by proton loss from the methyl group attached at C5 position. In parenthesis below each relative energy, delocalization length is indicated by counting the number of C atoms over which the spin is delocalized. For proton loss at a methyl group (attached at positions C5, C9 and C13 and by symmetry at C5′, C9′ and C13′), the number of carbon atoms was counted from the methyl group in the direction of delocalization. Note that the length over which the unpaired spin density is distributed increases as the relative energy decreases.

Car	$\Delta E(3)$	$\Delta E(4)$	$\Delta E(5)$	$\Delta E(9)$	$\Delta E(13)$	$\Delta E(13')$	$\Delta E(9')$	$\Delta E(6')$	$\Delta E(5')$	$\Delta E(4')$	$\Delta E(3')$	Reference
Astaxanthin	**0**/4.69	—	13.74	14.03	15.89	15.89	14.03	—	13.74	—	**0**/4.69	9
	(26/23)		(24)	(19)	(16)	(16)	(19)		(24)		(26/23)	
Zeaxanthin	—	**0**	3.15	8.39	10.21	10.21	8.39	—	3.15	**0**	—	2–4
		(23)	(23)	(19)	(15)	(19)	(15)		(23)	(23)		
β-Carotene	—	**0**	4.92	10.29	12.18	12.18	10.29	—	4.92	**0**	—	1,2,4
		(23)	(23)	(19)	(15)	(15)	(19)		(23)	(23)		
Lycopene	12.16	**0**	5.28	11.01	11.33	11.33	11.01		5.28	**0**	12.16	2,4,8
	(4)	(23)	(23)	(19)	(15)	(15)	(19)		(23)	(23)	(4)	
8′-Apo-β-caroten-8′-al	—	**0**	4.97	10.43	12.84	12.38	10.55	—	—	—	—	2,4
		(21)	(21)	(17)	(13)	(15)	(19)					
7′-Apo-7,7′-dicyano-β-carotene	—	**0**	5.34	11.11	12.85	13.99	13.81	—	—	—	—	2.4
		(23)	(23)	(19)	(15)							

Canthaxanthin	—	—	**0**	4.32	5.83	5.83	4.32	—	**0**	-	—	2,4
			(24)	(19)	(15)	(15)	(19)		(24)			
Lutein	—	6.68	6.70	15.16	17.06	17.04	14.72	**0**	22.69	45.44	—	2,4
		(21)	(21)	(17)	(13)	(15)	(19)	(23)	(4)	(4)		
Violaxanthin	—	15.33	19.43	**0**	1.78	1.78	**0**	—	19.4	15.33	—	2–4
		(4)	(4)	(17)	(13)	(13)	(17)		(4)	(4)		
9′-*cis*-Neoxanthin	—	—	25.6	2.6	5.0	5.1	**0**	—	22.5	—	—	5
			(4)	(17)	(14)	(14)	(17)		(4)			

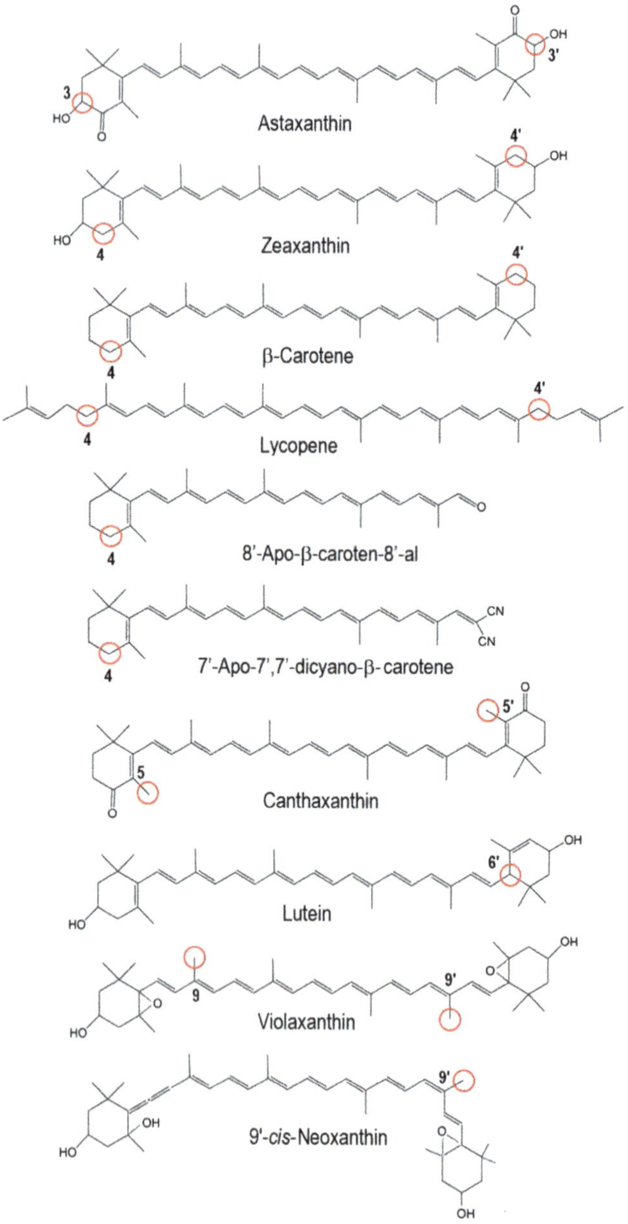
Astaxanthin
3
3'
HO
OH
Zeaxanthin
4
4'
β-Carotene
Lycopene
8'-Apo-β-caroten-8'-al
7'-Apo-7',7'-dicyano-β-carotene
CN
Canthaxanthin
5
5'
Lutein
6'
Violaxanthin
9
9'
9'-cis-Neoxanthin

Figure 3.6. Carotenoids studied and positions for the most favorable proton loss. Losing a proton from position(s) at the terminal end(s) of the radical cation forms neutral radicals that extend the conjugated system, with the spin delocalized over the whole π-conjugated system.

$\Delta E(3) = 4.69$ kcal/mol and delocalization length 23) over loss from the methyl groups of the polyene chain.[9]

For zeaxanthin radical cation $Zea^{\bullet+}$ the preferred proton loss occurs from the C4- or C4′-methylene positions on the cyclohexene rings (see Figure 3.7, zeaxanthin) to form $^{\#}Zea^{\bullet}(4)$, or by symmetry $^{\#}Zea(4')$ neutral radical.[3] The next energetically favorable positions are those by loss of the methyl proton attached at the C5(or C5′), C9(or C9′) or C13(or C13′) positions with relative energies $\Delta E = 3.15$, 8.39 and 10.21 kcal/mol, respectively (see relative energies for zeaxanthin neutral radicals in Table 3.10). We use notations $^{\#}Zea^{\bullet}(5)$ or $^{\#}Zea^{\bullet}(5')$, $^{\#}Zea^{\bullet}(9)$ or $^{\#}Zea^{\bullet}(9')$, $^{\#}Zea^{\bullet}(13)$ or $^{\#}Zea^{\bullet}(13')$ for the specific neutral radicals formed, proton loss position being indicated in parenthesis. The unpaired spin density for the carotenoid neutral radicals increases along the chain distant from the position from where the proton was lost (see Figure 3.7 on the left). The unpaired spin density distribution is larger for proton loss at C4(4′)-methylene position of the radical cation and decreases for proton loss from C5(5′)- to C9(9′)- to C13(13′)-methyl group, respectively. Thus, $^{\#}Zea^{\bullet}(4)$, the most favorable neutral radical of zeaxanthin (lowest energy structure), has a longer conjugation length, followed by $^{\#}Zea^{\bullet}(5)$, $^{\#}Zea^{\bullet}(9)$ and $^{\#}Zea^{\bullet}(13)$. In contrast, for violaxanthin the presence of the epoxy groups localizes the unpaired spin density on the C4(4′) and C5(5′) atoms respectively resulting in neutral radical structures that are energetically unfavorable.[3] Proton abstraction from C4(4′) position or C5(5′)-methyl group in violaxanthin requires very high energy, 15 to 20 kcal/mol higher than the most favorable structure formed by proton loss at C9(9′)-methyl group situated in the polyene chain (Table 3.10). These neutral radicals have never been observed by EPR measurements of violaxanthin supported on molecular sieves even though large 16 MHz couplings would be easily observed in a powder EPR spectrum.[3] The most favorable neutral

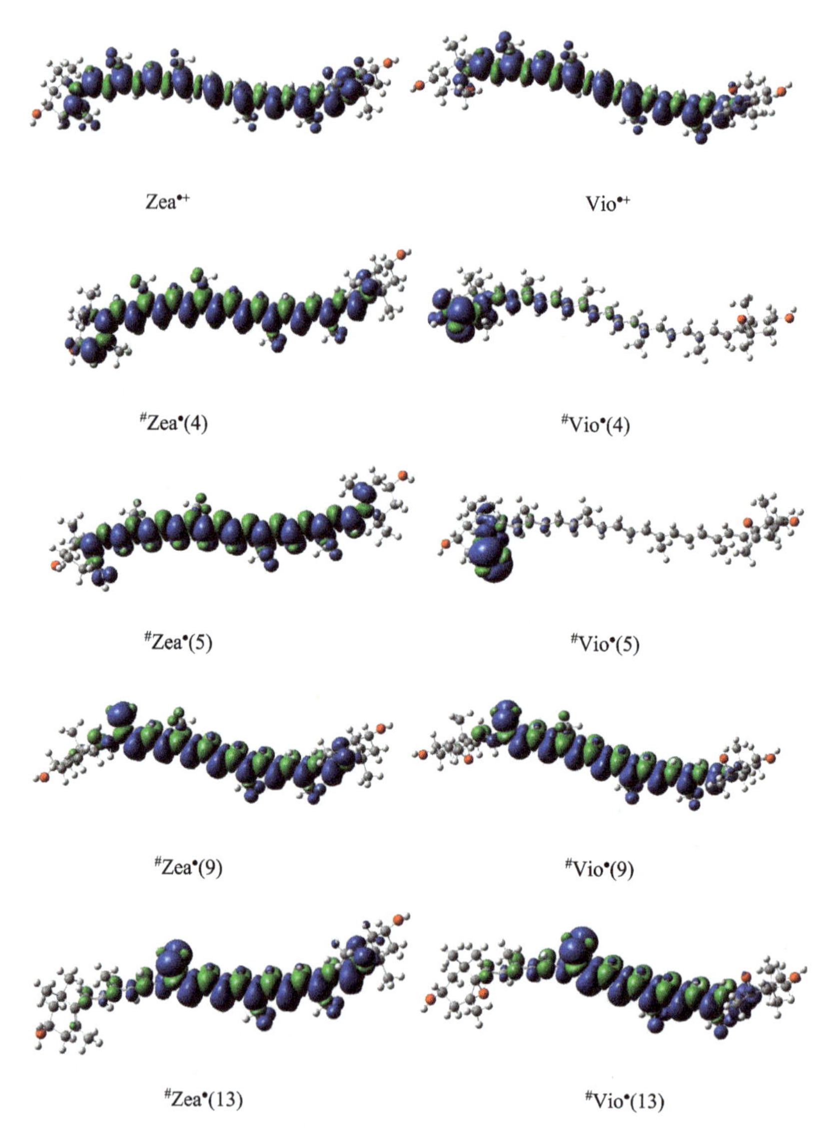

Figure 3.7. Unpaired spin distribution for zeaxanthin (left) and violaxanthin (right) radicals. Left, from top: Zea•+, #Zea•(4) (lowest energy structure, proton loss from terminal ends at C4(4´)-methylene positions), #Zea•(5), #Zea•(9), #Zea•(13). Right, from top: Vio•+, #Vio•(4) and #Vio•(5) (high-energy structures, not observed by EPR), #Vio•(9) (lowest-energy structure, proton loss from C9(9´)-methyl positions on the chain) and #Vio•(13). The blue represents excess α and the green excess β unpaired spin density. Adapted with permission from Reference 2.

radical for violaxanthin is that formed by loss of a methyl proton attached at C9(9′) position which extends conjugation the most (see Figure 3.7 on the right). Similarly to violaxanthin, for 9′-*cis*-neoxanthin, the most favorable proton loss occurs on the chain at C9′-methyl group followed by proton loss at C9-methyl group, rather than at the terminal rings (see Figure 3.7 and Table 3.10). Carotenoid 9′-*cis*-neoxanthin does not form neutral radicals by proton loss at C4(4′) or C5(5′) positions situated on the terminal rings (see high relative energies and short conjugation length in Table 3.10). The epoxy group at one terminal ring and an allene bond at the opposite end of the molecule disrupt the conjugation and do not allow fully conjugated neutral radicals to be formed.

For asymmetric carotenoids with just one terminal ring like 8′-apo-β-caroten-8′-al and 7′-apo-7,7′-dicyano-β-carotene, proton loss occurs from the C4-methylene position, followed by C5-methyl position on the ring, and ultimately followed by the methyl positions on the polyene chain.

For canthaxanthin, which has a carbonyl group on position C4 and symmetrically at C4′, the most favorable proton loss from the radical cation becomes that of a methyl proton from C5 or C5′, respectively. For asymmetric lutein, which is an isomer of zeaxanthin differing by a shift in a double bond by one position to C4′-C5′, the favored proton loss from the radical cation of lutein is no longer position C4 but position C6′, which is 6.68 kcal/mol lower than proton loss at C4, at the opposite end.[4] For the symmetric linear open chain lycopene,[8] proton loss from the radical cation to form the neutral radical is most stable at C4, or symmetrically at C4′, similar to the C4(4′)-methylene groups in β-carotene and zeaxanthin. For asymmetric 7′-apo-7,7′-dicyano-β-carotene and 8′-apo-β-caroten-8′-al, proton loss occurs from the cyclohexene ring at C4-methylene position followed by proton loss at the C5-methyl group (see Figure 3.7 and Table 3.10).

3.6 Conclusions

We have concluded that the unpaired spin density for the carotenoid neutral radicals increases at carbons along the chain distant from the position where the proton was lost. The longer the conjugation length, the more distributed the unpaired spin density and the more stable the radical. On the contrary, localized unpaired spin density gives rise to unstable radicals that are not observed in the EPR spectra.

Using hyperfine couplings generated at the B3LYP/TZP//B3LYP/6-31G** level, it was possible to show that carotenoids on a solid support, in the absence of light, produced carotenoid radical cations by electron transfer to the matrix. However, upon light irradiation (350–550 nm) the carotenoid neutral radicals are formed by proton loss from positions of the cyclohexene ring. The π-conjugated neutral radicals have the unpaired spin density distributed over the polyene chain. The identities of the radical cation and neutral radicals were confirmed through the determination of their isotropic β-methyl proton and anisotropic α-proton coupling constants. DFT-predicted β-methyl hyperfine coupling constants of 8 to 10 MHz of $Car^{\bullet+}$ were in good agreement with previously reported hyperfine couplings for carotenoid π-conjugated radicals with unpaired spin density delocalized over the whole molecule. Larger couplings of 13–16 MHz determined by DFT and present in ENDOR spectra of photo-irradiated carotenoids on solid supports or incorporated into MCM-41 molecular sieves were attributed to the formation of the neutral radicals.

References

1. Gao, Y., Focsan, A. L., Kispert, L. D., & Dixon, D. A. (2006). Density functional theory study of the beta-carotene radical cation and deprotonated radicals. *The Journal of Physical Chemistry B, 110*(48), 24750–24756. https://doi.org/10.1021/jp0643707

2. Focsan, A. L. (2008). EPR and DFT studies of proton loss from carotenoid radical cations. PhD Dissertation, The University of Alabama, Tuscaloosa, AL.
3. Focsan, A. L., Bowman, M. K., Konovalova, T. A., Molnár, P., Deli, J., Dixon, D. A., & Kispert, L. D. (2008). Pulsed EPR and DFT characterization of radicals produced by photo-oxidation of zeaxanthin and violaxanthin on silica-alumina. *The Journal of Physical Chemistry B, 112*(6), 1806–1819. https://doi.org/10.1021/jp0765650
4. Lawrence, J., Focsan, A. L., Konovalova, T. A., Molnar, P., Deli, J., Bowman, M. K., & Kispert, L. D. (2008). Pulsed electron nuclear double resonance studies of carotenoid oxidation in Cu(II)-substituted MCM-41 molecular sieves. *The Journal of Physical Chemistry B, 112*(17), 5449–5457. https://doi.org/10.1021/jp711310u
5. Focsan, A. L., Molnár, P., Deli, J., & Kispert, L. (2009). Structure and properties of 9′-cis neoxanthin carotenoid radicals by electron paramagnetic resonance measurements and density functional theory calculations: Present in LHC II? *The Journal of Physical Chemistry B, 113*(17), 6087–6096. https://doi.org/10.1021/jp810604s
6. Gao, Y., Shinopoulos, K. E., Tracewell, C. A., Focsan, A. L., Brudvig, G. W., & Kispert, L. D. (2009). Formation of carotenoid neutral radicals in photosystem II. *The Journal of Physical Chemistry B, 113*(29), 9901–9908. https://doi.org/10.1021/jp8075832
7. Polyakov, N. E., Focsan, A. L., Bowman, M. K., & Kispert, L. D. (2010). Free radical formation in novel carotenoid metal ion complexes of astaxanthin. *The Journal of Physical Chemistry B, 114*(50), 16968–16977. https://doi.org/10.1021/jp109039v
8. Focsan, A. L., Bowman, M. K., Molnár, P., Deli, J., & Kispert, L. D. (2011). Carotenoid radical formation: Dependence on conjugation length. *The Journal of Physical Chemistry B, 115*(30), 9495–9506. https://doi.org/10.1021/jp204787b
9. Focsan, A. L., Bowman, M. K., Shamshina, J., Krzyaniak, M. D., Magyar, A., Polyakov, N. E., & Kispert, L. D. (2012). EPR study of the astaxanthin n-octanoic acid monoester and diester radicals on silica-alumina. *The Journal of Physical Chemistry B, 116*(44), 13200–13210. https://doi.org/10.1021/jp307421e

10. Focsan, A. L., Magyar, A., & Kispert, L. D. (2015). Chemistry of carotenoid neutral radicals. *Archives of Biochemistry and Biophysics, 572*, 167–174. https://doi.org/10.1016/j.abb.2015.02.005
11. Focsan, A. L., & Kispert, L. D. (2017). Radicals formed from proton loss of carotenoid radical cations: A special form of carotenoid neutral radical occurring in photoprotection. *Journal of Photochemistry and Photobiology B: Biology, 166*, 148–157. https://doi.org/10.1016/j.jphotobiol.2016.11.015
12. Focsan, A. L., Polyakov, N. E., & Kispert, L. D. (2017). Photo protection of *Haematococcus pluvialis* algae by astaxanthin: Unique properties of astaxanthin deduced by EPR, optical and electrochemical studies. *Antioxidants, 6*(4), 80. https://doi.org/10.3390/antiox6040080
13. Gao, Y., Focsan, A. L., & Kispert, L. D. (2020). Antioxidant activity in supramolecular carotenoid complexes favored by nonpolar environment and disfavored by hydrogen bonding. *Antioxidants, 9*(7), 625. https://doi.org/10.3390/antiox9070625
14. Gao, Y., Focsan, A. L., & Kispert, L. D. (2020). The effect of polarity of environment on the antioxidant activity of carotenoids. *Chemical Physics Letters, 761*(16), 138098. https://doi.org/10.1016/j.cplett.2020.138098
15. Focsan, A. L., Polyakov, N. E., & Kispert, L. D. (2021). Carotenoids: Importance in daily life-insight gained from EPR and ENDOR. *Applied Magnetic Resonance, 52*(8), 1093–1112. https://doi.org/10.1007/s00723-021-01311-8
16. Frisch, M. J., Trucks, G. W., Schlegel, H. B., Scuseria, G. E., Robb, M. A., Cheeseman, J. R., Montgomery, J. A., Jr., Vreven, T., Kudin, K. N., Burant, J. C., Millam, J. M., Iyengar, S. S., Tomasi, J., Barone, V., Mennucci, B., Cossi, M., Scalmani, G., Rega, N., Petersson, G. A., ... Pople, J. A. (2003). *Gaussian 03*, Revision B.02; Gaussian, Inc.: Pittsburgh, PA.
17. Cho, H., Felmy, A. R., Craciun, R., Keenum, J. P., Shah, N., & Dixon, D. A. (2006). Solution state structure determination of silicate oligomers by 29SI NMR spectroscopy and molecular modeling. *Journal of the American Chemical Society, 128*(7), 2324–2335. https://doi.org/10.1021/ja0559202
18. Filatov, M., & Cremer, D. (2005). Calculation of spin-densities within the context of density functional theory. The crucial role of the correlation functional. *The Journal of Physical Chemistry, 123*, 124101. https://doi.org/10.1063/1.2047467

19. Tomasi, J., Mennucci, B., & Cammi, R. (2005). Quantum mechanical continuum solvation models. *Chemical Reviews, 105*(8), 2999–3094. https://doi.org/10.1021/cr9904009
20. Faller, P., Maly, T., Rutherford, A. W., & MacMillan, F. (2001). Chlorophyll and carotenoid radicals in photosystem II studied by pulsed ENDOR. *Biochemistry, 40*, 320–326.
21. Himo, F. (2001). Density functional theory study of the β-carotene radical cation. *The Journal of Physical Chemistry A, 105*, 7933–7937. https://doi.org/10.1021/jp011473a
22. Wu, Y., Piekara-Sady, L., & Kispert, L. D. (1991). Photochemically generated carotenoid radicals on Nafion film and silica gel: An EPR and ENDOR study. *Chemical Physics Letters, 180*, 573–577. https://doi.org/10.1016/0009-2614(91)85012-L
23. Piekara-Sady, L., Khaled, M. M., Bradford, E., Kispert, L. D., & Plato, M. (1991). Comparison of the INDO to the RHF-INDO/SP derived EPR proton hyperfine couplings for the carotenoid cation radical: Experimental evidence. *Chemical Physics Letters, 186*, 143–148. https://doi.org/10.1016/S0009-2614(91)85120-L
24. Jeevarajan, A. S., Kispert, L. D., & Piekara-Sady, L. (1993). An ENDOR study of carotenoid cation radicals on silica-alumina solid supports. *Chemical Physics Letters, 209*(3), 269–274. https://doi.org/10.1016/0009-2614(93)80106-Y
25. (a) Becke, A. D. (1993). Density-functional thermochemistry. III. The role of exact exchange. *The Journal of Chemical Physics, 98*(7), 5648–5652. http://dx.doi.org/10.1063/1.464913 (b) Lee, C., Yang, W., & Parr, R. G. (1988). Development of the Colle-Salvetti correlation-energy formula into a functional of the electron density. *Physical Review B, 37*(2), 785–789. https://doi.org/10.1103/PhysRevB.37.785
26. Hehre, W. J., Radom, L., Schleyer, P. R., & Pople, J. A. (1986). *Ab initio Molecular Orbital Theory*, John Wiley and Sons.
27. Schafer, A., Horn, H., & Ahlrichs, R. (1992). Fully optimized contracted Gaussian basis sets for atoms Li to Kr. *The Journal of Chemical Physics, 97*(4), 2571–2577. https://doi.org/10.1063/1.463096

28. Grant, J. L., Kramer, V. J., Ding, R., & Kispert, L. D. (1988). Carotenoid cation radicals: Electrochemical, optical, and EPR study. *Journal of the American Chemical Society, 110*(7), 2151–2157. https://doi.org/10.1021/ja00215a025
29. Jeevarajan, A. S., Khaled, M., & Kispert, L. D. (1994). Simultaneous electrochemical and electron paramagnetic resonance studies of keto and hydroxy carotenoids. *Chemical Physics Letters, 225*, 340–345. https://doi.org/10.1016/0009-2614(94)87091-8
30. Khaled, M. (1992). Electrochemical, transient EPR, and AM1 excited states studies of carotenoids. PhD dissertation, The University of Alabama, Tuscaloosa, AL.
31. Jeevarajan, A. S., Khaled, M., & Kispert, L. D. (1994). Simultaneous electrochemical and electron paramagnetic resonance studies of carotenoids: Effect of electron donating and accepting substituents. *The Journal of Physical Chemistry, 98*(32), 7777–7781. https://doi.org/10.1021/j100083a006
32. Gao, Y., Webb, S., & Kispert, L. D. (2003). Deprotonation of carotenoid radical cation and formation of a didehydrodimer. *The Journal of Physical Chemistry B, 107*, 13237–13240. https://doi.org/10.1021/jp0358679
33. Dixon, D. A., & Arduengo, A. J. III (2006). Accurate heats of formation of the "Arduengo-type" carbene and various adducts including H2 from ab initio molecular orbital theory. *The Journal of Physical Chemistry A, 110*, 1968–1974. https://doi.org/10.1021/jp055527i
34. Magill, A. M., Cavell, K. J., & Yates, B. F. (2004). Basicity of nucleophilic carbenes in aqueous and nonaqueous solvents-Theoretical predictions. *Journal of the American Chemical Society, 126*(28), 8717–8724. https://doi.org/10.1021/ja038973x
35. Magill, A. M., & Yates, B. F. (2004). An assessment of theoretical protocols for calculation of the pK_a values of the prototype imidazolium cation. *Australian Journal of Chemistry, 57*(12), 1205–1210. https://doi.org/10.1071/CH04159
36. Alexeev, Y., Windus, T. L., Zhan, C.-G., & Dixon, D. A. (2005). Accurate heats of formation and acidities for H_3PO_4, H_2SO_4, and H_2CO_3 from ab initio electronic structure calculations. *International Journal of Quantum Chemistry, 102*, 775–784. Erratum. *International Journal of Quantum Chemistry, 104*, 379–380.

4

Electron Paramagnetic Resonance Spectroscopy Measurements

Starting in the late 1980s, for more than 20 years in Prof. Kispert's lab at The University of Alabama in Tuscaloosa, carotenoid radicals have been detected and their properties elucidated using Electron Paramagnetic Resonance (EPR) spectroscopy techniques.[1–51]

In Table 4.1, the various EPR techniques with references used to study the carotenoid radicals are listed. Some of these measurements have been outlined in Chapter 9 of the book *Carotenoids*[47] and are revised herein.

4.1 EPR of Electrochemically Formed Carotenoid Radical Cations

In 1988, for the first time, Grant *et al.*[1] detected using EPR carotenoid radical cations generated electrochemically in solution. EPR spectra of 10^{-3} M solutions of three different carotenoids in tetrahydrofuran, methylene chloride, and dichloroethane were recorded following *in situ* electrochemical oxidation. There were no radicals detected in tetrahydrofuran but in the chlorinated solvents all three carotenoids gave rise to an unresolved EPR single Gaussian line with g_{iso} around 2.0026, indicating that the radical cation was produced. In each case, the expected hyperfine structure was not observed, and each line was relatively broad, approximately 13–14 Gauss for β-carotene and canthaxanthin, and 18–20 Gauss for 8′-apo-β-caroten-8′-al. The X band (9 GHz) EPR spectrum of a carotenoid radical cation consists of an unresolved, inhomogeneously broadened line with $g = 2.0027 \pm 0.0002$, characteristic of organic π radicals (Figure 4.1).

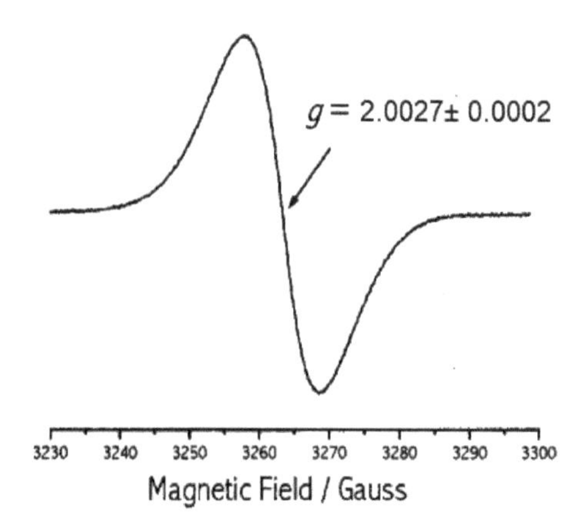

Figure 4.1. EPR (9 GHz) spectrum of a carotenoid radical cation. Adapted with permission from Figure 1 of Reference 25.

4.2 EPR of Chemically Formed Carotenoid Radical Cations

Carotenoid radical cations were also formed in reaction with iodine and detected using EPR.[2] Reaction between I_2 and carotenoids formed the $I_3^-\cdots Car^{\bullet+}$ complex. At 77 K, the equilibrium shifts so $Car^{\bullet+}\ldots I_n^-$ formed, where n = 5, 7 or 9. The I_n polyene chain resided over the carotenoid polyene chain in a π-π interaction which was detected as a measurable shift in the g-value. Another EPR study has shown that one-electron transfer reactions occur between carotenoids and quinones.[23] A charge-transfer complex is formed in equilibrium with an ion-radical pair ($Car^{\bullet+}\ldots Q^{\bullet-}$). Increasing the temperature from 77 K gives rise to the quinone radical anions. At room temperature a carotenoid-quinone radical adduct is formed.[23] As a parenthesis, oxidation of carotenoids by ferric ion is also possible, shown in our studies (other than EPR) by CV and simultaneous optical absorption spectroscopy[52] or by ^{1}H-NMR, LC-MS and optical measurements.[53] Depending on the concentration of Fe^{3+}, Fe^{2+} and Cl^- relative to the carotenoid concentration, $Car^{\bullet+}$ or Car^{2+} can be formed.[53] The radical cation and dication was also found to abstract an electron from Fe^{2+}.

Isomerization occurs during oxidation. In the presence of air, ^{1}H-NMR, LC-MS, and optical studies have shown that a stable product is formed in high yield, the 5,8-peroxide of the starting carotenoid.[53]

4.3 Simultaneous Electrochemical and EPR (SEEPR) Measurements

This method allowed the simultaneous recording of the cyclic voltammetry (CV) and EPR spectrum of the carotenoid radicals during electron transfer reactions.[4,13,14] All these SEEPR experiments were carried out with an IBM-enhanced electrolytic cell, ER/164 ECA, using a Varian E-12 EPR spectrometer and a V-4533 rotating cylindrical cavity. The special electrolyte cell was inserted in the rotating cylindrical EPR cavity which permitted the CVs to be measured by a commercial electrochemical analyzer while simultaneously recording the EPR spectrum. During the electrochemical oxidation of canthaxanthin in solution,[4] a strong EPR signal for the radical cation was observed since comproportionation constant K_{com} for canthaxanthin was large, on the order of 10^3. Carotenoids with strong electron acceptor substituents like canthaxanthin exhibit large K_{com}, while carotenoids with electron donor substituents like β-carotene exhibit very small K_{com} which resulted in formation of diamagnetic carotenoid dications in solution rather than EPR-detectable radical cations. This explains the strong EPR spectrum observed for canthaxanthin and the weak EPR spectrum observed for β-carotene (Figure 4.2).

SEEPR measurements have also shown that carotenoids substituted with electron-donating groups are more easily oxidized than those with electron-accepting substituents.[13] Comproportionation constants for the dication and the neutral species were measured by SEEPR techniques and by simulation of the CVs. The width of the SEEPR spectrum of the radical cation for carotenoids with electron-donating groups varied from 13.2 to 15.6 G, and

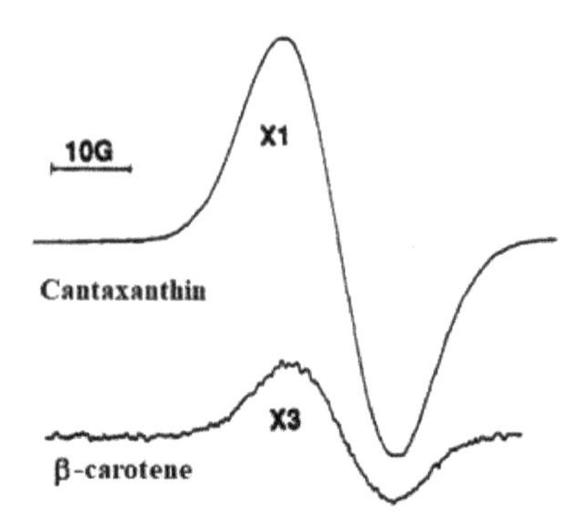

Figure 4.2. EPR spectra of canthaxanthin and β-carotene. Adapted with permission from Figure 8 of Reference 4.

the g-factors were 2.0027 ± 0.0002.[13] Comproportionation constants from EPR spin concentration were also deduced for keto and hydroxy carotenoids that were electrochemically oxidized. The line widths of the SEEPR spectra of the radical cations were in the range of 13.2 to 14.5 G and the g-factors were 2.0027 ± 0.0002, thus confirming the polyene π-radical cation structure.[14]

Simultaneous EPR and electrochemical measurements[1,3,4] also gave information about radical lifetimes. Carotenoid radical cations are formed electrochemically in CH_2Cl_2 solvent with a half-life of ~0.2 minutes and exhibit a decay half-life of 1–2 minutes. Upon deuteration of the carotenoid, these half-lives increase by an order of magnitude.[3] The half-lives of radical cations in dichloromethane formed electrochemically and chemically with ferric chloride were followed optically and by stopped flow, and were found to be dependent on the number of conjugated chain double bonds.

4.4 Continuous Wave Electron Nuclear Double Resonance (CW ENDOR) to Detect β-methyl Protons

A carotenoid radical contains many different proton hyperfine couplings (~18) which results in approximately 300,000 unresolved EPR lines for symmetrical carotenoids, and an even larger number of unresolved EPR lines for asymmetrical carotenoids. Fortunately, the rapid rotation of the methyl group protons averages out the proton dipolar anisotropy giving rise to only one set of proton couplings. Even with reducing the number of couplings due to methyl groups rotation, there is still a very large number of proton couplings that results in one unresolved, inhomogeneous broadened powder EPR line of 14 Gauss peak-to-peak linewidth.

To deduce more information from the EPR spectrum, a double resonance technique called electron nuclear double resonance (ENDOR) was used. To resolve the hyperfine couplings of carotenoid radicals, continuous wave (CW) ENDOR measurements have been carried out starting in the early 1990s[5–8,15] until around year 2006, after which pulsed ENDOR was more predominantly used (see Table 4.1). Details of the CW ENDOR method, equipment operation and examples of spectral analysis have been

Table 4.1. Various EPR techniques used to study carotenoid radicals.

Technique	References
EPR of electrochemically formed carotenoid radical cations in solution	1, 3
EPR of chemically formed carotenoid radical cations	2, 23
Simultaneous electrochemical and EPR (SEEPR) detection of the transient radical prepared in solution	4, 13, 14
CW ENDOR to detect β-methyl protons	5–8, 11, 12, 15, 19, 25–28, 30, 36, 37, 46
Time-resolved EPR measurements	9, 10, 16
Photoinduced electron transfer in frozen solutions	18, 19
High-frequency/high magnetic field EPR to resolve g-anisotropy	21
High-frequency/high magnetic field EPR to identify high-spin metal centers	26, 28

(Continued)

Table 4.1. (*Continued*)

Technique	References
EPR studies on solid supports	5, 25–29, 31, 32
EPR spin trapping	22–24
EPR studies of carotenoid complexes	33, 35, 40, 44, 50
ESEEM methods and pulsed EPR relaxation techniques for measuring distances between carotenoid radicals and distant metals	25, 34, 42
HYSCORE to detect α-proton anisotropic couplings	25, 38
Pulsed ENDOR to detect β-methyl proton couplings	25, 38, 39, 41, 46, 48, 49
DFT calculations to interpret the powder ENDOR and HYSCORE spectra	36, 38, 39, 41, 46, 48, 49

published by Goslar *et al.* (1994),[11] Piekara-Sady and Kispert (1994),[12] Kispert and Piekara-Sady (2006)[37] and Keavan and Kispert (1976).[54] In an ENDOR spectrum, instead of multiple EPR lines for each set of equivalent protons, only two ENDOR lines separated by the hyperfine coupling constant, A, occur.

For best resolution, ENDOR measurements should be carried out in solution where proton dipolar anisotropy is averaged out. However, the carotenoid radical cations are short-lived in chlorinating electron acceptor solvents with lifetimes on the order of 10–200 s, and are even shorter in other solvents.[3] Thus, a steady state mM concentration needed for ENDOR measurements in a sample tube lasting 30 minutes to 1 hour could not be achieved. Producing the radicals electrochemically *in situ* is not possible as the radiofrequency from the ENDOR instrument would heat up the samples to boiling. Only one ENDOR experiment by electrolyzing a sample of canthaxanthin external to the cavity was successful (Figure 4.3).[7] The radical cation of canthaxanthin generated electrochemically in dichloromethane and its solution ENDOR spectrum was reported for the first time in 1993.

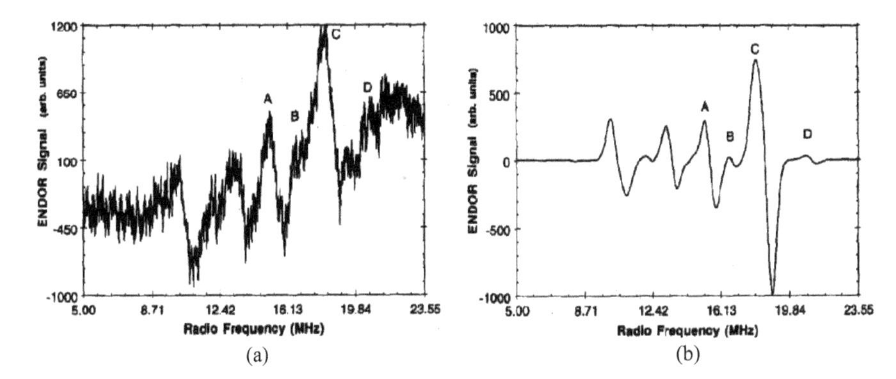

Figure 4.3. ENDOR spectrum of the electrochemically prepared cation radical of canthaxanthin at 210 K: (a) experimental and (b) simulated. Adapted with permission from Figure 1 of Reference 7.

This was possible because canthaxanthin has a large K_{com} which lies to the right toward the formation of radical cations, and the measurement was enabled by a steady state concentration for 30 min.[7] By contrast, for β-carotene, with low K_{com} that lies to the left favoring formation of the diamagnetic dication instead, the EPR signal of β-carotene radical cation in solution appears too weak for an ENDOR measurement. The intense ENDOR spectral lines that appear in frozen carotenoid solution or carotenoids adsorbed on powders are due to proton couplings between 0.3 and 8.3 MHz assignable to the β-methyl protons attached to the carbons C13, C9, and C5, as well as the α-proton couplings.[6,8] Notably, the proton couplings for the methyl groups at C13, C9, and C5 positions are resolved due to rapid rotation of the methyl groups averaging out any dipolar anisotropy which would normally have broadened the ENDOR line beyond detection.

As discussed in Chapter 3, the INDO-calculated proton couplings of carotenoid radical cations[5,6,8] were found to be overestimated. Density Functional Theory (DFT) calculations have shown[36] that the error was sufficient that the formation of the carotenoid neutral radical via the loss of a proton from the carotenoid radical cation was missed in earlier studies. Carotenoid neutral radicals are formed upon photoirradiation and are detectable by the observation of an ENDOR line at 21–23 MHz. Since the methyl groups rapidly rotate relative to the microfrequency even at 5 and 670 GHz,[21] the resolved ENDOR spectra of carotenoid radicals in frozen solutions could be observed, similar to those of carotenoid radicals on Nafion film,[5] silica alumina[8] and silica gel.[15] The resolved spectra were due to couplings from the protons on the rotating methyl groups. Further studies[19] showed that the electron is transferred from the carotenoid molecules adsorbed on the activated alumina or silica alumina to the surface Lewis acid sites since ^{27}Al couplings were detected.

Adsorption of carotenoids on solid supports such as activated silica alumina results in the formation of radical cations by electron transfer between carotenoid molecules and the Al^{3+} electron acceptor.[25] Subsequent loss of a proton from the weak acid radical cation forms the neutral radicals. The three-pulse electron-spin echo envelope modulation (ESEEM) spectrum and the hyperfine sublevel correlation (HYSCORE) spectrum of canthaxanthin in methylene chloride containing $AlCl_3$ show a peak at ^{27}Al Larmor frequency (3.75 MHz),[25] proving the existence of electron transfer between Al^{3+} Lewis acid sites and carotenoids on silica alumina. This electron transfer results in the formation and stabilization of carotenoid radical cations. Tumbling of carotenoid molecules adsorbed on the solid support is restricted, but methyl groups can rotate, increasing the lifetime of the radicals to days from the usual 100 seconds in solution. This rotation is the only type of dynamic processes which is evident in the CW ENDOR spectra.[19,25] High-field 330 GHz EPR,[21] ESEEM and Davies-pulsed ENDOR and HYSCORE measurements[25] have proven useful in further examining the structures and features of the adsorbed carotenoid radicals on silica alumina. Davies-pulsed ENDOR and 330 GHz EPR measurements from 3.3 to 80 K showed no line-shape changes, implying very rapid rotation of the methyl groups down to 3.3 K.

CW ENDOR was used to distinguish the spectrum of the radical cation $Car^{\bullet +}$ from that of the neutral radical $^{\#}Car^{\bullet}$. For that, carotenoids were adsorbed on catalysts like silica alumina, MCM-41 or metal-substituted MCM-41 where they would form radicals that were stable for hours.[51] Based on the calculated DFT hyperfine couplings it was determined that a mixture of radical cations $Car^{\bullet +}$ and neutral radicals $^{\#}Car^{\bullet}$ was formed. Any line above 19 MHz was assigned to the neutral radicals $^{\#}Car^{\bullet}$, below that value peaks indicated a mixture of $Car^{\bullet +}$ and $^{\#}Car^{\bullet}$. For example, the simulated and experimental spectrum for an irradiated sample of β-carotene on silica alumina showed contribution from both β-$Car^{\bullet +}$ and $^{\#}$β-$Car^{\bullet}$ (see Figure 4.4, also discussed in a previous chapter).[36] In a non-irradiated solid sample the lines were assigned using DFT-calculated couplings to

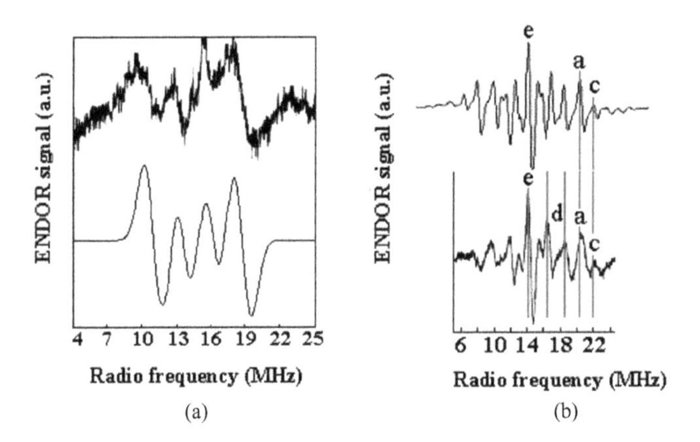

Figure 4.4. (a) Experimental powder ENDOR spectrum for the radical cation of β-carotene prepared on a silica alumina matrix in the absence of UV irradiation (top) and simulated powder ENDOR spectrum (bottom) assuming isotropic proton couplings of 8.3, 7.0 and 2.6 MHz (linewidth 1.5 MHz); only the rapid averaging of the methyl proton couplings can be detected by powder ENDOR measurements, the ENDOR lines for the anisotropic α-proton couplings are too broad to detect. Adapted with permission from Reference 8. (b) Simulation of CW ENDOR of β-carotene radicals (top) and experimental in the presence of UV irradiation (bottom). Lines above 19 MHz are due to neutral radicals. Adapted with permission from References 5 and 36.

the radical cation $Car^{•+}$ only (Figure 4.4a).[8] The typical isotropic hyperfine coupling for a carotenoid radical cation is around 8 MHz, and there are no lines above 19 MHz. The light irradiation of the samples and the presence of metals in the solid matrix produces more neutral radicals $^{\#}Car^{•}$, with larger values of the isotropic hyperfine couplings, larger than 13 MHz, as deduced from the ENDOR lines at approximately 22 MHz (Figure 4.4b).

Different studies have shown that the concentration of radicals formed on solid supports depends on the type of carotenoid (Figure 4.5a),[28] seems to be higher in the presence of a metal (Figure 4.5b, c, d)[26,27,29] and depends on the type of metal: for example, it was higher in the presence of Al than Ni (Figure 4.5c).[26]

4.5 Time-resolved EPR Measurements

Photoexcitation products generated by laser pulse photolysis (XeCl laser, 308 nm) of carotenoids in liquid CCl_4 at room temperature have been studied by X-band (9 GHz) and Q-band (35 GHz) time-resolved electron paramagnetic resonance (TREPR) spectroscopy. In the TREPR measurements[9,10,16] polarized chemically induced dynamic electron polarized (CIDEP) 35 GHz spectra[16] showed that for β-carotene dissolved in carbon tetrachloride (CCl_4), a solvent-separated radical ion pair between $CCl_4^{•-}$ and carotenoid radical cations $Car^{•+}$ lasted for several μs after the laser pulse (Figure 4.6). From the polarization pattern, it was concluded that the electron transfer occurs from the excited singlet state of β-carotene to the CCl_4 solvent.

4.6 Photoinduced Electron Transfer in Frozen Solutions

To lengthen the time available for an EPR study, carotenoid solutions in different electron acceptor solvents such as CH_2Cl_2, $CHCl_3$, CCl_4 and CS_2 were frozen at 77 K and the photoinduced electron transfer from

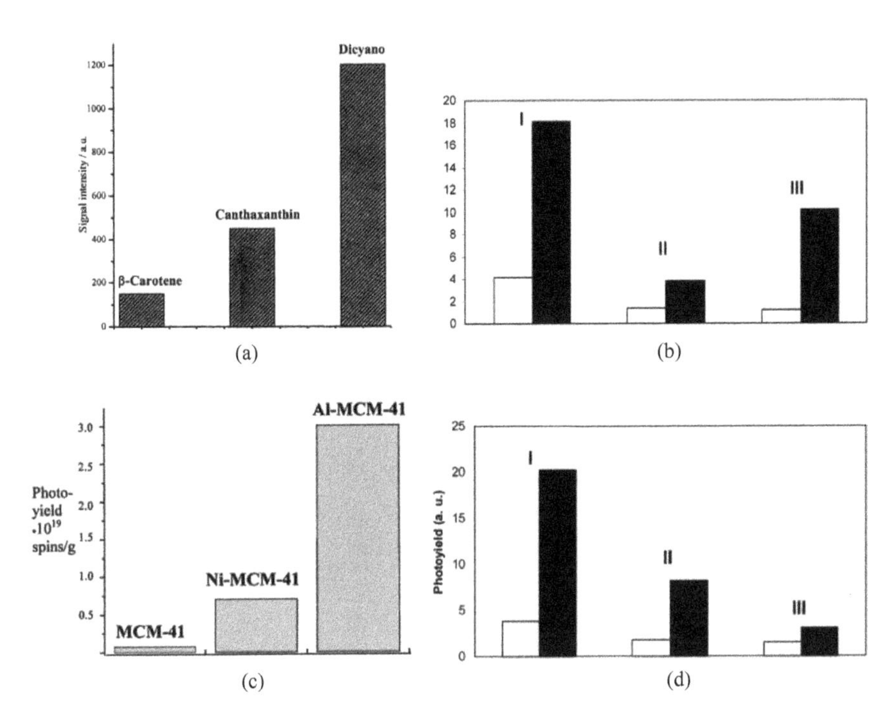

Figure 4.5. (a) Yield of carotenoid radical cations in Fe-MCM-41 molecular sieves. Adapted with permission from Figure 9 of Reference 28. (b) Photoyield comparison of β-carotene (I), canthaxanthin (II), and 7′-apo-7′,7′-dicyano-β-carotene (III) imbedded in MCM-41 and Ti-MCM-41. Blank columns represent photoyield in MCM-41; black columns represent that in Ti-MCM-41. Adapted with permission from Figure 5 of Reference 27. (c) The photoyield of carotenoid radical cations in MCM-41, Ni-MCM-41, and Al-MCM-41 molecular sieves. Adapted with permission from Figure 8 in Reference 26. (d) Photoyield comparison of β-carotene (I), 8′-Apo-β-caroten-8′-al (II) and canthaxanthin (III) imbedded in MCM-41 and Cu-MCM-41. Blank columns represent the photoyield in MCM-41; black columns represent that in Cu-MCM-41. Adapted with permission from Figure 9 in Reference 29.

carotenoids to solvent molecules was monitored.[18] In these media, long-lived solvent separated radical ion pairs were stable for several days and the yield was 10 times higher in chlorinated solvents than in CS_2. In the case of CS_2 solutions the $CS_2^{\bullet-}$ radical anions are observed at 77 K while at higher temperature, the radicals formed were converted to dimer radical

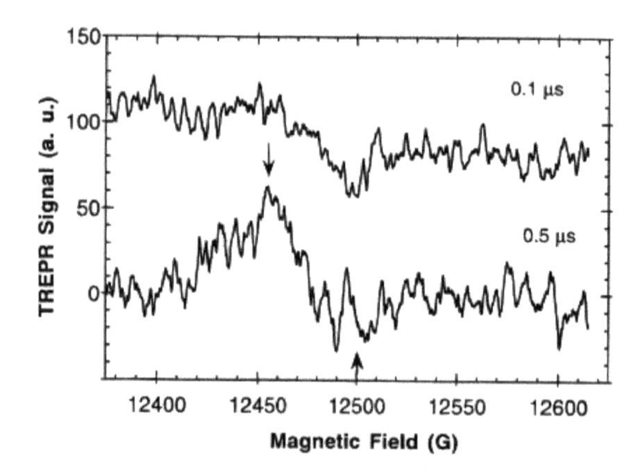

Figure 4.6. TREPR signal (35 GHz) after 0.1 (top) and 0.5 μs (bottom) delay of 308 nm laser light photolysis of β-carotene (0.2 mM) in CCl4 purged with nitrogen. Adapted with permission from Figure 4.1 in Reference 16.

anions $(CS_2)_2^{\bullet-}$. For chlorinated solvents the radicals were assigned to either a predissociation complex $(R\ldots Cl)^{\bullet-}$ or to radical products $R^{\bullet}$. This EPR study also enabled the study of peroxyl radicals ($RO_2^{\bullet}$) in oxygen-saturated carotenoid solution where the detailed mechanism has been established.[18]

Photolysis of β-carotene and canthaxanthin frozen solutions in CCl_4, $CHCl_3$, CH_2Cl_2, and CS_2 resulted in the formation of these stable paramagnetic species for several days at 77 K. A representative EPR spectrum is shown in Figure 4.7 for canthaxanthin in CH_2Cl_2. It was found that EPR signals are similar for both β-carotene and canthaxanthin carotenoid solutions. Warming of the samples to 250–270 K, even for only a second, caused immediate disappearance of all EPR signals, although radical cations of β-carotene and canthaxanthin generated electrochemically in CH_2Cl_2 are known to be stable at room temperature for several minutes. The warming effect can be explained by recombination of the photogenerated radical pairs, which may be the carotenoid radical cations $Car^{\bullet+}$ and solvent counter radical anions $Sol^{\bullet-}$ or their reaction products. Photolysis of neat solvents under the same conditions did not produce any EPR signals.

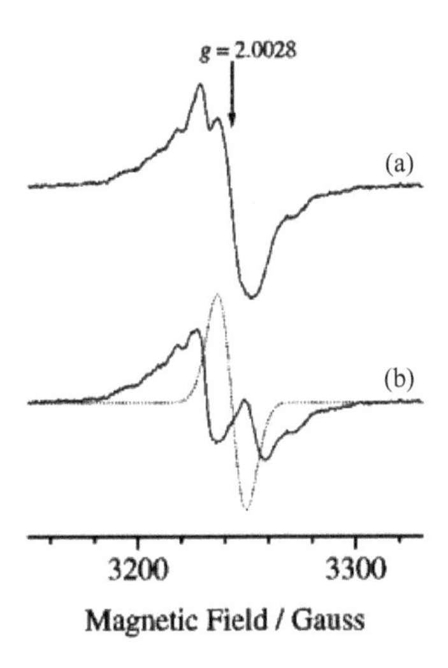

Figure 4.7. EPR spectra of degassed canthaxanthin solution in CH_2Cl_2 at 77 K: (a) irradiation by Xe/Hg lamp (578 nm) at 77 K; (b) (solid line) difference spectrum after subtraction of $Car^{\bullet+}$ signal, (dotted line) electrochemically generated at 298 K, then cooled to 77 K and measured at that temperature. Adapted with permission from Figure 1 of Reference 18.

4.7 High-frequency/high-field EPR for g-tensor Resolution

High-frequency/high-field EPR (HFEPR) spectroscopy greatly improves the resolution of the EPR signals. These measurements were used to demonstrate that the symmetrical unresolved EPR line at 9 GHz is due to a carotenoid π-radical cation with electron density distributed throughout the entire chain as predicted by molecular orbital calculations.[21] For example, the canthaxanthin radical cation formed by adsorbing canthaxanthin on silica alumina gives rise to an unresolved EPR single Gaussian line with $g_{iso} = 2.0027$ characteristic of an organic π-radical. However, at higher frequencies from 327 to 670 GHz, the unresolved line resolves into two peaks as a result of g anisotropy of $g_{\perp} = 2.0023$ and $g_{\parallel} = 2.0032$, characteristic of a cylindrically symmetrical π-radical cation (Figure 4.8). The difference $g_{xx} - g_{yy}$ decreases with increasing chain length. When $g_{xx} - g_{yy}$

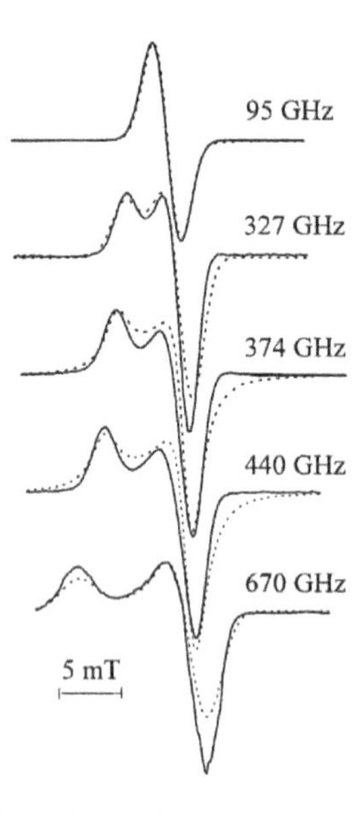

Figure 4.8. HFEPR spectra of canthaxanthin. Adapted with permission from Figure 1 of Reference 21.

approaches zero, the g-tensor becomes cylindrically symmetrical with $g_{xx} = g_{yy} = g_{\perp}$ and $g_{zz} = g_{\parallel}$ for the all-*trans* carotenoid radical cation. Determination of the g-tensor components from resolved 327–670 GHz EPR spectra allowed differentiation between carotenoid radical cations with cylindrical symmetry from other C–H-containing π-radicals of different symmetry (Table 4.2). Also, the lack of temperature dependence of the EPR line widths over the range of 5–80 K at 327 GHz suggested rapid rotation of methyl groups even at 5 K that averages out the proton couplings from three oriented β-methyl protons.

4.8 High-frequency/high-field EPR Measurements of Metal Centers

HFEPR is also a promising technique to increase spectral resolution for proper assignment of different metal ion sites, which cannot be resolved by the X-band experiments. HFEPR can be used to good advantage to study high-spin systems, where the zero-field splitting term is often dominant in

Table 4.2. Comparison of g-values for various radical cations with those observed for canthaxanthin radical cation. Adapted with permission from Reference 21 and references therein.

Radical cations	g_{xx}	g_{yy}	g_{zz}	g_{iso}	Structure	Reference
$TP^{\bullet+}/AsF_5^-$	2.0031	2.0028	2.0022	2.0027	Polymer	Robinson *et al.* (1985)
$TP^{\bullet+}/PF_6^-$	2.00312	2.00230	2.00206	2.00249	Stacked array	Kispert *et al.* (1987)
$QP^{\bullet+}/PF_6^-$	2.00217	2.00217	2.00310	2.00248	Stacked array	Kispert *et al.* (1987)
$P865^{\bullet+}$	2.00337	2.00248	2.00208	2.00264	Dimer	Klette *et al.* (1993)
$P700^{\bullet+}$	2.00317	2.00260	2.00226	2.0027	Dimer	Bratt *et al.* (1997)
$Bchla^{\bullet+}$	2.0033	2.0026	2.0022	2.0027	Monomer	Burghaus *et al.* (1991)
$Car^{\bullet+}$	2.0023	2.0023	2.0032	2.0026	Symmetrical π-radical cation	Konovalova *et al.* (1999)

the spin Hamiltonian. The examples of such systems are transition metal ions like Mn^{2+}, Ni^{2+}, and Fe^{3+} which have been used for introducing active sites in mobile crystalline mesoporous materials like MCM-41. MCM-41 containing well-organized nanometer-sized channels has been found to be a good photoredox system where long-lived photoinduced electron transfer from bulky biomolecules such as carotenoids can occur. Although the MCM-41 framework can act as an electron acceptor, replacement of some tetrahedral Si(IV) in the MCM-41 framework by metal ions such as Ni,[26] Al[26] or Fe[28] produces a long-lived charge separation between the carotenoid radicals and the metal electron acceptor sites.

Photo-oxidation of β-carotene in mesoporous Ni-MCM-41 molecular sieves was studied by 9-220 GHz HFEPR spectroscopy.[26] The presence

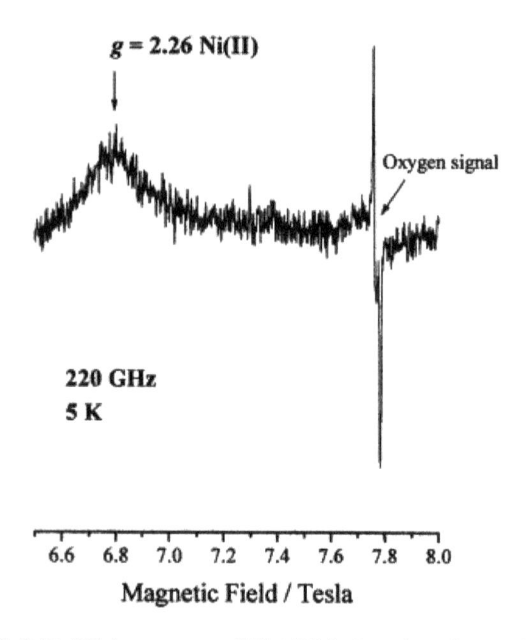

Figure 4.9. EPR (220 GHz) spectrum of Ni-MCM-41 activated at 260°C, degassed, and measured at 5 K. Adapted with permission from Figure 3 of Reference 26.

of Ni^{2+} ions in Ni-MCM-41 was verified by 220 GHz EPR spectroscopy because samples measured at 9 GHz showed no EPR signals consistent with Ni^{2+} ions. The 220 GHz EPR spectrum of activated Ni-MCM-41 measured at 5 K exhibits a broad line with g-value of 2.26 providing direct evidence of Ni^{2+} incorporation into the MCM-41 framework (Figure 4.9). This is in agreement with data reported for Ni^{2+} ions in an octahedral environment that give rise to very broad EPR lines with g-values of 2.10–2.33.

After 350 nm irradiation of the Ni-MCM-41 sample, the EPR (110 GHz) spectrum at 5 K showed two new paramagnetic species stable at 77 K: one was assigned to the O_2^- species and the other to V-centers or trapped holes on the framework oxygens generated in MCM-41 (Figure 4.10).

Photooxidation of carotenoids in Ni-MCM-41 produced an intense EPR signal with g-value of 2.0027 due to the formation of carotenoid

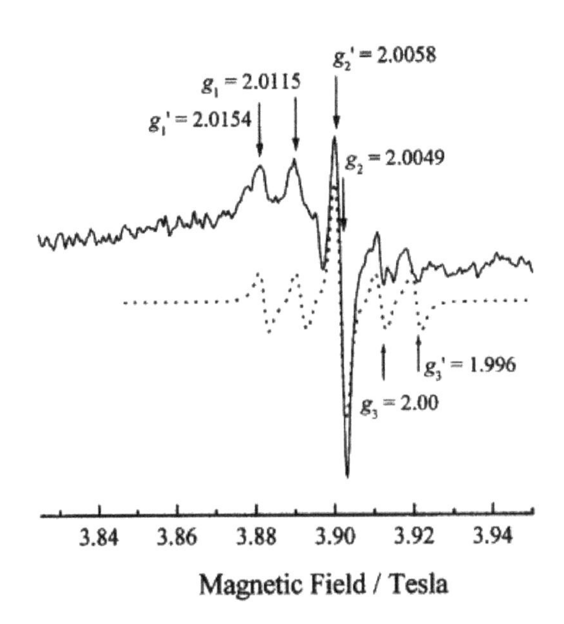

Figure 4.10. EPR (110 GHz) spectrum at 5 K of Ni-MCM-41 after 350 nm irradiation (solid line) and simulated spectrum (dotted line). The signal with a rhombic g tensor g_1 = 2.0115, g_2 = 2.0049, g_3 = 2.00 was assigned to O_2^- species generated in MCM-41. The second rhombic g tensor $g_1{'}$ = 2.0154, $g_2{'}$ = 2.0058, $g_3{'}$ = 1.996 was assigned to V-centers. Adapted with permission from Figure 4 of Reference 26.

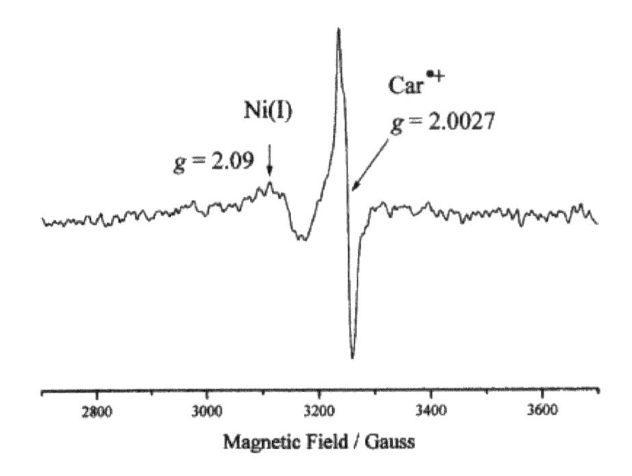

Figure 4.11. EPR (9 GHz) spectrum at 77 K of β-carotene in Ni-MCM-41 after 350 nm irradiation. Adapted with permission from Figure 5 of Reference 26.

radical cations and the other less intense line with g = 2.09 was attributed to a Ni^{+} species produced as a result of electron transfer from the carotenoid molecule to Ni^{2+} (Figure 4.11). The Ni(I) EPR signals were not detected upon 350 nm irradiation of Ni-MCM-41 samples before adsorption of carotenoid, so the irradiated sample containing the carotenoid provided direct evidence for reduction of Ni^{2+} ions by carotenoids.

HFEPR 9-287 GHz was also used to characterize Fe^{3+} sites in Fe-MCM-41 molecular sieves.[28] Spectra characteristic of framework high-spin Fe^{3+} and activating Fe-MCM-41 at higher temperature diminished the g = 4.3 framework iron signal and significantly increases the extra-framework iron signal at 2.0 which demonstrates that the tetrahedral coordination of framework Fe^{3+} is not very stable. It was found that high-frequency/high-field EPR increased spectral resolution for proper assignment of different Fe^{3+} sites that are not resolved at 9 GHz. Measurements at 9 GHz cannot identify which one of the iron sites can react with the carotenoid.

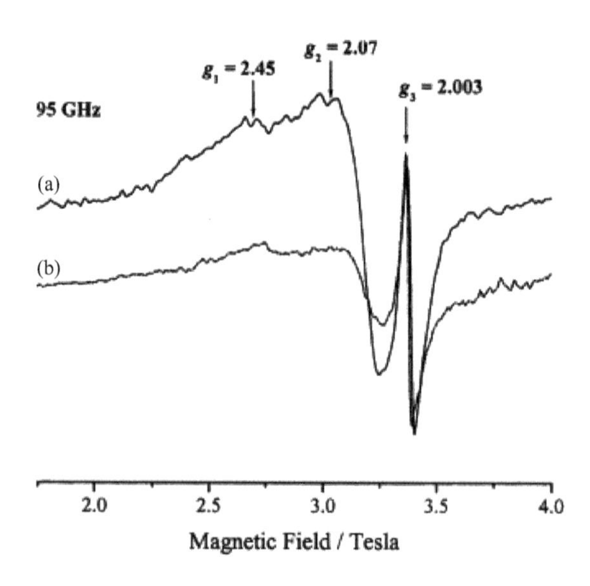

Figure 4.12. EPR (95 GHz) spectra of Fe-MCM-41 (a) in the absence of carotenoid; (b) after adsorption of canthaxanthin. Adapted with permission from Figure 7 of Reference 28.

However, at 95 GHz, it was shown that the extra-framework Fe^{3+} ion located on the surface of the pore is primarily responsible for carotenoid oxidation — possibly these sites are more accessible for bulky organic molecules than the framework iron within silica walls. Further studies, when carotenoid was incorporated in Fe-MCM-41 and subjected to irradiation at 365 nm for 2 min, showed signals for the radical cation and V-centers of the framework oxygens (Figure 4.12). Since V centers are formed during the reduction of metal ions, it was suggested that oxidation of carotenoids in Fe-MCM-41 proceeds through electron transfer from carotenoid molecules to the electron acceptor sites producing Fe^{2+}-$O^{\bullet}$-Si V-centers.

4.9 HFEPR: Effect of Distant Metals on g-tensor

When an organic radical is located near a high-spin metal ion, the g-tensor of the radical depends on the exchange interaction between the radical and the metal ion. Multi-frequency HFEPR measurements are needed to determine precise g-values of the exchange-coupled organic radical. Details were presented on the procedure.[47] g-tensor anisotropy variation was also shown to be sensitive to the presences or absence of dimer or multimer-stacked structures. It was shown that dimers that occur for radical cations can be deduced by monitoring the g_{yy} component.

4.10 EPR Studies on Solid Supports

Carotenoid cation radicals have been observed at 120 K, by EPR and proton ENDOR measurements, to be formed upon 77 K photolysis of Nafion thin films or silica gel coated with the carotenoids.[5] We have studied the oxidation of carotenoids adsorbed on silica alumina or imbedded in the pores of MCM-41, or metal-substituted MCM-41 where their radical cation is formed. At X-band EPR frequency, the radical cation exhibits a single unresolved peak with g_{iso} = 2.0026 characteristic of organic π-radicals.[1] Additionally, in most of these studies ENDOR was performed for detection of the hyperfine couplings.[5,25–28]

4.11 EPR Spin Trapping of Carotenoids

Combining spin-trapping EPR technique and UV-vis spectroscopy[22] has determined the relative rates of reaction of carotenoids with the $^{\bullet}OOH$ radical formed by the Fenton reaction in DMSO solvent.

$$Fe^{2+} + H_2O_2 \rightarrow Fe^{3+} + {}^{\bullet}OH + OH^-$$
$${}^{\bullet}OH + DMSO \rightarrow {}^{\bullet}CH_3 + CH_3SO(OH)$$
$$H_2O_2 + {}^{\bullet}CH_3 \rightarrow {}^{\bullet}OOH + CH_4$$

A very non-linear increase in trapping rate of carotenoids with $^{\bullet}OOH$ occurs as the oxidation potential of the carotenoid increases (Figure 4.1). Carotenoids with high oxidation potential like canthaxanthin (0.775 V vs. SCE) and astaxanthin (0.768 V vs. SCE) are excellent scavengers of the $^{\bullet}OOH$ radical by abstraction of the most acidic proton of the carotenoid, showing an antioxidant activity. However, for carotenoids with low oxidation potential like β-carotene (0.634 V vs. SCE) and zeaxanthin (0.616 V vs. SCE), a pro-oxidant activity occurs where the Fe^{3+} ion formed in the Fenton reaction oxidizes the carotenoid molecule Car to form the radical cation $Car^{\bullet+}$ (Scheme 4.1). This results in more radicals being formed, a pro-oxidant activity.

In water solution xanthophylls like zeaxanthin and lutein form H-aggregates which significantly reduces the scavenging rate.[50] In fact, zeaxanthin exhibits a pro-oxidant effect while lutein which possesses a slightly larger oxidation potential exhibits little or no pro-oxidant effect. This may be due to the greater proton-donating rate of the unsymmetrical lutein aggregates compared to the symmetrical zeaxanthin aggregates. Astaxanthin also forms H-aggregates (side by side dimer-like). Aggregation in

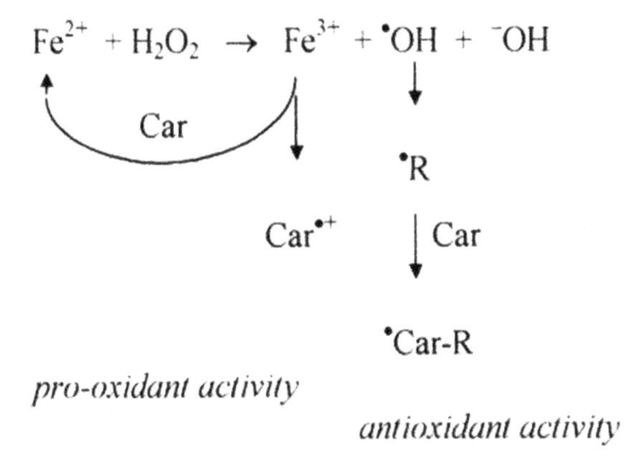

Scheme 4.1. The formation of the radical cation $Car^{\bullet +}$ is the pro-oxidant activity and •Car-R is the antioxidant activity (adapted with permission from Reference 24).

water solution is a specific feature of hydroxy-containing carotenoids. The substitution of hydroxyl groups of astaxanthin to ester groups inhibits aggregation completely.[50]

4.12 EPR Studies of Carotenoid Complexes

Water-soluble supramolecular complexes of carotenoids have been formed with oligosaccharides cyclodextrin (CD)[33] and glycyrrhizic acid (GA),[35,50] as well as with polysaccharide arabinogalactan (AG),[40,44] resulting in an increase in water solubility of the normally hydrophobic carotenoids. These complexes are discussed in more detail in another chapter. Here we mention those EPR experiments that helped elucidate their structures and properties.

EPR spin trapping along with ^{1}H-NMR and optical studies on inclusion complexes of carotenoids with CD showed that CD protects the carotenoid from reactive oxygen species but complexation results in considerable decrease in antioxidant ability of the carotenoid.[33] By using the EPR spin-trapping technique, the scavenging ability of the carotenoid toward •OOH radicals was compared in DMSO and in the aqueous CD solution.

A considerable decrease in PBN/•OOH spin adduct yield was detected in the presence of uncomplexed carotenoid because of a competing reaction of the carotenoid with •OOH radical. No such decrease occurred in the presence of the CD complex with the carotenoid. Moreover, the small increase in spin adduct yield for the complex indicates a pro-oxidant effect, most likely due to the reaction of the carotenoid with Fe^{3+} to regenerate Fe^{2+}, which in turn regenerates the •OOH radical.[33]

Complexes of carotenoid with AG resulted in increased photostability[40] and TiO_2 photocatalytic activity.[44] EPR demonstrated the increased stability of the carotenoid radical cation imbedded into AG host, similar to that of carotenoid radical cations in GA.[40] EPR spin trapping showed that the carotenoid-AG complex provides the maximum photocatalytic efficiency during visible light irradiation with $\lambda > 380$ nm. As compared with pure carotenoids, complexes with AG exhibit an enhanced quantum yield of free radicals and stability toward photodegradation.[44] These results are important in the design of artificial light-harvesting, photoredox, and catalytic devices.

Xanthophyll carotenoids can self-assemble in aqueous solution to form J- and H-type aggregates.[50] When carotenoids lie side by side, an H-type aggregate occurs and it is detected by a blue shift (~100 nm shift). J-aggregates form when carotenoids lie tail to head and are optically detected by a red shift. We have applied EPR spin trapping and optical absorption spectroscopy to investigate[50] how complexation with AG and GA can affect the aggregation ability of the xanthophyll carotenoids zeaxanthin, lutein, and astaxanthin, as well as their photostability and antioxidant activity. Complexation with both AG and GA reduced the aggregation rate but did not inhibit aggregation completely. Inclusion complexes of AG and GA with both monomer and H-aggregates of carotenoids were formed. H-aggregates of carotenoids exhibit higher photostability in aqueous solutions as compared with monomers, but much lower antioxidant activity. Complexation increases the photostability of both monomers and the aggregates of xanthophyll carotenoids. In the presence of water, lutein shows much lower antioxidant activity due to the formation of H-aggregates. Under the same conditions, zeaxanthin shows a pro-oxidant effect in the Fenton reaction. Complexes of zeaxanthin, lutein and astaxanthin with GA demonstrate higher antioxidant activity in aqueous solution as GA forms a donut-like dimer in which the hydrophobic polyene chain of the xanthophylls and their H-aggregates lie protected within the donut hole, permitting the hydrophilic ends and most acidic proton to be exposed to the surroundings.[50]

4.13 Electron Spin-Echo Envelope Modulation

The interaction of carotenoid radicals with their surroundings can be deduced by applying advanced EPR methods. For example, ESEEM (three-pulse) measurements[34] have been used to determine the distance-dependent reversible electron transfer[29] from a deuterated carotenoid to the host lattice Cu-MCM-41-containing Cu^{2+} (Figure 4.13). ESEEM and pulse

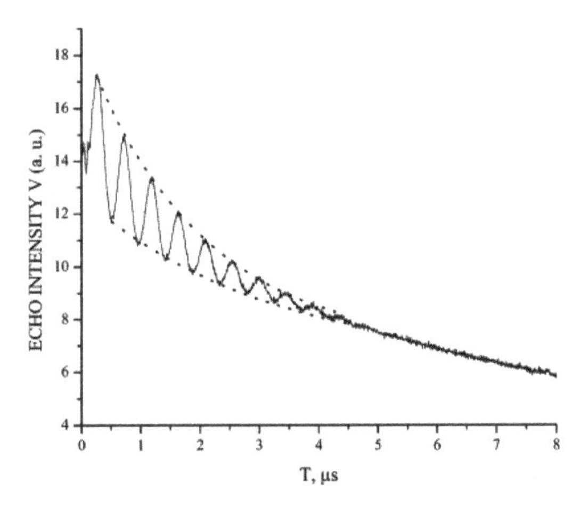

Figure 4.13. Three-pulse ESEEM of Cu-MCM-41 with adsorbed β-carotene. Adapted with permission from Figure 2 of Reference 34.

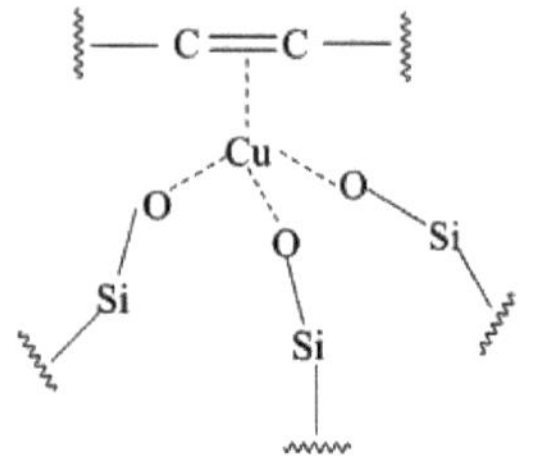

Scheme 4.2. Cu^{2+}-Car complex in Cu-MCM-41. Adapted with permission from Reference 29.

ENDOR studies[34] of Cu^{2+}-Car complex in Cu-MCM-41 showed that Cu^{2+} interacts with the centrally located double bond C15=C15′. Cu^{2+} was found to reside 2.8 Å from the center of the polyene chain. The short distance between Cu^{2+} and carotenoid in the Cu^{+2}-Car complex favors light-driven electron transfer from Car to Cu^{2+} and also permits the thermal back electron transfer rate from Cu^{+} to $Car^{\bullet+}$ (Scheme 4.2).[29]

4.14 Pulsed EPR Relaxation Enhancement Measurements

Details have been published on how to use the enhancement in carotenoid relaxation rates caused by a metal ion to determine the distance between a metal framework ion and the carotenoid radical cation formed.[42] From these measurements it can be determined how far an electron can be transferred to the framework to form a carotenoid radical cation. In the published example,[42] the effect of a rapidly relaxing framework Ti^{3+} ion on spin lattice relaxation time T_1 and phase memory time T_M of a slowly relaxing carotenoid radical were measured as a function of temperature in both siliceous and Ti-substituted MCM-41. The phase memory times T_M of the carotenoid radicals were determined from the best fits of two-pulse ESEEM curves. The spin-lattice relaxation times T_1 of the Ti^{3+} were obtained from the inversion recovery experiment with echo detection on a logarithmic time scale in the temperature range of 10–150 K. T_M and

T_1 are shorter for carotenoids embedded in Ti-MCM-41 than in siliceous MCM-41. The most dramatic effect on T_M occurs when the metal relaxation rate is of the same order of magnitude as the dipolar couplings to the slowly relaxing spin. It was found[42] that the interspin distances between the carotenoid radical and Ti^{3+} were 13.0 ± 3.0 Å for canthaxanthin (and there was no interaction with Ti^{3+}), 13.0 ± 2.0 Å for a carotenoid containing a carboxyl group (which would interact strongly with Ti sites) and 9.0 ± 3.0 Å for the short-chain polyene, β-ionone.[42]

4.15 α-Protons from HYSCORE Analysis of Powder Spectra

HYSCORE is a 2D four-pulse ESEEM technique which provides correlation between nuclei frequencies from different electron manifolds and from which the anisotropic hyperfine couplings can be deduced.[25] The α-proton anisotropic couplings were detected by this technique. α-proton hyperfine couplings of carotenoid radicals can be determined from 2D-HYSCORE analysis of the contour line-shapes of the cross-peaks which provide the principal components of the tensors. The tensors for the radicals of zeaxanthin and violaxanthin photo-generated on silica alumina were rhombic, characteristic of planar conjugated radicals with the unpaired spin in a p_z orbital of the carbon of the C–H group.[38] Details of the analysis procedure have been summarized in Reference 47. On the basis of our DFT calculations, the hyperfine couplings determined from the HYSCORE analysis of the photo-oxidation of zeaxanthin adsorbed on silica alumina were assigned to α-protons of the radical cation and neutral radicals formed by deprotonation (Figure 4.14).[38] By comparison with the DFT-calculated anisotropic couplings we confirmed the presence of the radical cation and neutral radicals for zeaxanthin and the absence of the violaxanthin neutral radicals formed by proton loss at the terminal rings, just as deduced from the ENDOR measurements.[38]

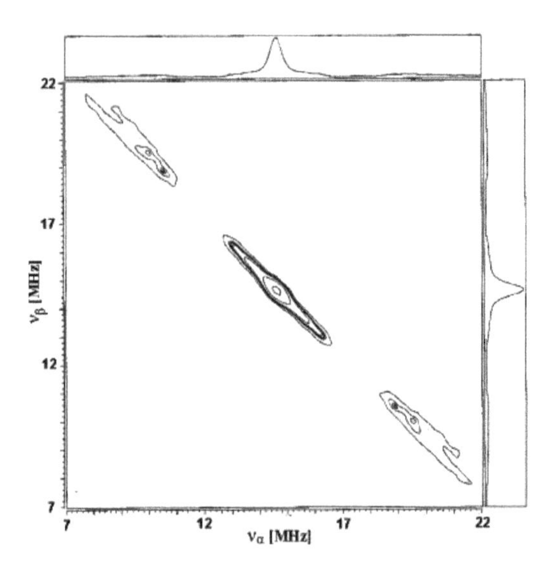

Figure 4.14. Contour plot of the HYSCORE spectrum of zeaxanthin radicals adsorbed on silica alumina. Parameters: T = 40 K, B = 3422 G, ν = 9.672145 GHz, τ = 172 ns, repetition time = 799.68 μs. Adapted with permission from Figure 5 of Reference 38.

4.16 Pulsed ENDOR: Davies and Mims ENDOR to Detect β-proton Hyperfine Couplings

Pulsed Davies and Mims ENDOR were used to detect and distinguish carotenoid neutral radicals from the radical cations that are formed upon photo-irradiation of carotenoids adsorbed on silica alumina,[25,36,38,46,48,49] the molecular sieves MCM-41,[41,46] Ti-MCM-41[42,46] or Cu-MCM-41.[39,43] Davies- and Mims-pulsed ENDOR methods complement each other because hyperfine couplings are present in one method but not in the other due to different pulse delay times. Examples have been published for β-carotene radicals on silica alumina[36] and in Cu-MCM-41,[43] zeaxanthin and violaxanthin on silica alumina,[38] lutein radicals formed in Cu-MCM-41 matrices,[39] 9′-*cis*-neoxanthin radicals in MCM-41,[41] lycopene in Cu-MCM-41[39] and on activated silica alumina,[48] and astaxanthin and its esters on silica alumina, MCM-41 and Ti-MCM-41.[46,49]

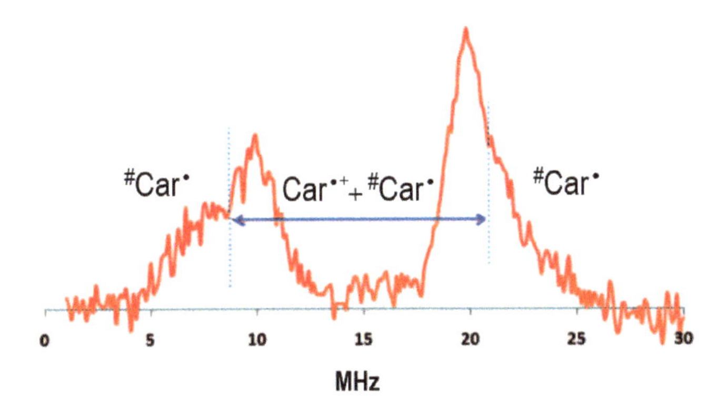

Figure 4.15. Experimental Davies ENDOR spectrum of irradiated 9′-*cis*-neoxanthin radicals on MCM-41 at 20 K. Adapted with permission from Figure 3 of Reference 41.

Davies spectra of carotenoids indicated that a mixture of radical cations and neutral radicals was formed[38,41] but due to the poor resolution (Figure 4.15) we turned to Mims ENDOR as a complementary method to study carotenoids.[39,41,43,48,49,51] Using this method, by varying the delay time (τ) which is the time between the first and second pulses in the Mims ENDOR experiment, features in the spectrum can be deleted where the ENDOR amplitude goes to zero causing a blind spot. The ENDOR amplitude goes to zero according to $1 - \cos(2\pi A\tau)$ where A is the hyperfine coupling, so having DFT-calculated hyperfine couplings and selecting different delay times can erase or show the presence of certain peaks. This was instrumental in demonstrating that peaks corresponding to the carotenoid neutral radicals appear at the edges of the Mims spectrum at high and low frequency, above 20 MHz and below 10 MHz in Figure 4.16. In other words, by varying the delay time we could map out the neutral radicals differentiating them from the radical cation.[51]

4.17 DFT Calculations used to Interpret ENDOR Spectra

DFT calculations of the minimum energy for proton loss and structures of various carotenoid neutral radicals have been carried out for numerous

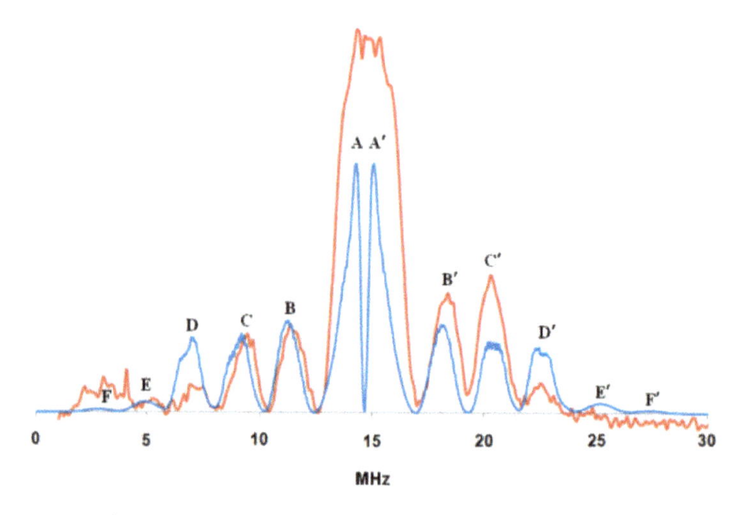

Figure 4.16. Mims ENDOR spectra of irradiated 9′-*cis*-neoxanthin radicals in MCM-41. Red: experimental spectrum measured at T = 20 K. Blue: simulated spectrum using DFT proton hyperfine coupling tensors for radical cation and neutral radicals. Adapted with permission from Figure 4 of Reference 41.

carotenoids and ENDOR was used for detection of the neutral radicals stabilized on the various matrices as summarized in Table 3.9 in the previous chapter. DFT calculations[36] were used starting in 2006 to correct the hyperfine couplings[5,6,8,15,17,26] for the carotenoid radical cation, misassigned earlier by the only available INDO or AM1 calculations. By using different DFT functionals and basis sets to calculate hyperfine couplings that fit the experimental couplings, it was confirmed that the isotropic β-methyl proton hyperfine couplings do not exceed 9–10 MHz for the carotenoid radical cations. DFT calculations for carotenoid neutral radicals formed by proton loss of the most acidic proton from the radical cation were shown to exhibit isotopic β-methyl couplings up to 16 MHz.[36,38,46,48,49] This explained the large isotopic couplings previously observed by CW ENDOR measurements of irradiated carotenoids supported on silica gel, Nafion films, silica alumina matrices or incorporated in molecular sieves.[5,6,15,19,25,26]

The DFT calculations showed that the most energetically favorable neutral radical formed by deprotonation from the radical cation at positions situated on the terminal groups that extend conjugation. As a consequence of deprotonation, a change in the unpaired spin distribution occurs which produces larger methyl proton hyperfine constants (13–16 MHz) than for the radical cation (8–10 MHz). Several papers have been published on the unpaired distribution that occurs for various carotenoids as a function of substituent, symmetry, and function.[36,38,39,41,46,48,49] The unpaired spin density for the carotenoid neutral radicals increases at carbons along the chain distant from the position from where the proton was lost. Proton loss prevented from the terminal rings results in structures that are energetically unfavorable and have never been observed by EPR measurements.[37]

References

1. Grant, J. L., Kramer, V. J., Ding, R., & Kispert, L. D. (1988). Carotenoid cation radicals: An electrochemical, optical, and EPR study. *Journal of the American Chemical Society, 110*, 2151–2157. https://doi.org/10.1021/ja00215a025
2. Ding, R., Grant, J. L., Metzger, R. J., & Kispert, L. D. (1988). Carotenoid cation radicals produced by the interaction of carotenoids with iodine. *The Journal of Physical Chemistry, 92*, 4600–4606. https://doi.org/10.1021/j100327a010
3. Khaled, M., Hadjipetrou, A., & Kispert, L. (1990). Electrochemical and electron paramagnetic resonance studies of carotenoid cation radicals and dications: Effect of deuteration. *The Journal of Physical Chemistry, 94*, 5164–5169. https://doi.org/10.1021/j100375a072
4. Khaled, M., Hadjipetrou, A., Kispert, L. D., & Allendoerfer, R. D. (1991). Simultaneous electrochemical and electron paramagnetic resonance studies of carotenoid cation radicals and dications. *The Journal of Physical Chemistry, 95*, 2438–2442. https://doi.org/10.1021/j100159a060
5. Wu, Y., Piekara-Sady, L., & Kispert, L. D. (1991). Photochemically generated carotenoid radicals on Nafion film and silica gel: An EPR and ENDOR

study. *Chemical Physics Letters, 180*, 573–577. https://doi.org/10.1016/0009-2614(91)85012-L

6. Piekara-Sady, L., Khaled, M. M., Bradford, E., Kispert L. D., & Plato, M. (1991). Comparison of the INDO to the RHF-INDO/SP derived EPR proton hyperfine couplings for the carotenoid cation radical: Experimental evidence. *Chemical Physics Letters, 186*, 143–148. https://doi.org/10.1016/S0009-2614(91)85120-L
7. Piekara-Sady, L., Jeevarajan, A. S., & Kispert, L. D. (1993). An ENDOR study of the canthaxanthin cation radical in solution. *Chemical Physics Letters, 207*, 173–177. https://doi.org/10.1016/0009-2614(93)87010-Z
8. Jeevarajan, A. S., Kispert, L. D., & Piekara-Sady, L. (1993). An ENDOR study of carotenoid cation radicals on silica alumina solid supports. *Chemical Physics Letters, 209*, 269–274. https://doi.org/10.1016/0009-2614(93)80106-Y
9. Jeevarajan, A. S., Khaled, M., Forbes, M. D. E., & Kispert, L. D. (1993). CIDEP studies of carotenoid radical cations. *Zeitschrift für Physikalische Chemie, 182*, 51–61. https://doi.org/10.1524/zpch.1993.182.Part_1_2.051
10. Jeevarajan, A. S., Khaled, M., Forbes, M. D. E., & Kispert, L. D. (1993). CIDEP studies of carotenoid radical cations. In: *Magnetic Field and Spin Effects in Chemistry*, Part II, R. Oldenbourg Verlag.
11. Goslar, J., Piekara-Sady, L., & Kispert, L. (1994). ENDOR data tables. In: C. P. Poole, Jr. & H. A. Farach (Eds.), *Handbook of Electron Spin Resonance: Data Sources, Computer Technology, Relaxation and ENDOR* (pp. 359–652). American Institute of Physics Press.
12. Piekara-Sady, L., & Kispert, L. D. (1994). ENDOR Spectroscopy. In: C. P. Poole, Jr. & H. A. Farach (Eds.), *Handbook of Electron Spin Resonance: Data Sources, Computer Technology, Relaxation and ENDOR* (pp. 311–358). American Institute of Physics Press.
13. Jeevarajan, A. S., Khaled, M., & Kispert, L. D. (1994). Simultaneous electrochemical and electron paramagnetic resonance studies of carotenoids: Effect of electron donating and accepting substituents. *The Journal of Physical Chemistry, 98*, 7777–7781. https://doi.org/10.1021/j100083a006
14. Jeevarajan, A. S., Khaled, M., & Kispert, L. D. (1994). Simultaneous electrochemical and electron paramagnetic resonance studies of keto and

hydroxy carotenoids. *Chemical Physics Letters, 225*, 340–345. https://doi.org/10.1016/0009-2614(94)87091-8

15. Piekara-Sady, L., Jeevarajan, A. S., Kispert, L. D., Bradford, E. G., & Plato, M. (1995). ENDOR study of the (7′,7′-dicyano)-and (7′-phenyl)-7′-apo-β-carotene radical cations formed by UV photolysis of carotenoids adsorbed on silica gel. *Journal of the Chemical Society, Faraday Transactions, 91*, 2881–2884. https://doi.org/10.1039/FT9959102881
16. Jeevarajan, A. S., Kispert, L. D., Avdievich, N. I., & Forbes, M. D. E. (1996). Role of excited singlet state in the photooxidation of carotenoids: A time resolved Q-band EPR study. *The Journal of Physical Chemistry, 100*, 669–671. https://doi.org/10.1021/jp952180i
17. Jeevarajan, J. A., Jeevarajan, A. S., & Kispert, L. D. (1996) Electrochemical, EPR and AM1 studies of acetylenic carotenoids. *Journal of the Chemical Society, Faraday Transactions, 92*, 1757–1765. https://doi.org/10.1039/FT9969201757
18. Konovalova, T. A., Konovalov, V. V., & Kispert, L. D. (1997), Photoinduced electron transfer between carotenoids and solvent molecules. *The Journal of Physical Chemistry, 101*, 7858–7862. https://doi.org/10.1021/jp9708761
19. Konovalova, T. A., & Kispert, L. D. (1998). EPR and ENDOR studies of carotenoid-solid Lewis acid interactions. *Journal of the Chemical Society, Faraday Transactions, 94*, 1465–1468. https://doi.org/10.1039/A709049H
20. Konovalova, T. A., Kispert, L. D., & Konovalov, V. V. (1999). Surface modification of TiO_2 nanoparticles with carotenoids: An EPR study. *The Journal of Physical Chemistry B, 103*, 4672–4677. https://doi.org/10.1021/jp9900638
21. Konovalova, T. A., Krzystek, J., Bratt, P. J., Van Tol, J., Brunel, L.-C., & Kispert, L. D. (1999). 95–670 GHz EPR studies of canthaxanthin radical cation stabilized on a silica-alumina surface. *The Journal of Physical Chemistry B, 103*, 5782–5786. https://doi.org/10.1021/jp990579r
22. Polyakov, N. E., Kruppa, A. I., Leshina, T. V., Konovalova, T. A., & Kispert, L. D. (2001). Carotenoids as antioxidants: Spin trapping EPR and optical study. *Free Radical Biology and Medicine, 31*(1), 43–52. https://doi.org/10.1016/s0891-5849(01)00547-0

23. Polyakov, N. E., Konovalov, V. V., Leshinia, T. V., Luzina, O. A., Salakhutdinov, N. F., Konovalova, T. A., & Kispert, L. D. (2001). One-electron transfer product of quinone addition to carotenoids EPR and optical absorption studies. *Journal of Photochemistry and Photobiology A: Chemistry, 141*, 117–126. https://doi.org/10.1016/S1010-6030(01)00429-4
24. Polyakov, N. E., Leshina, T. V., Konovalova, T. A., & Kispert, L. D. (2001). Carotenoids as scavengers of free radicals in a Fenton reaction: Antioxidants or pro-oxidants? *Free Radical Biology and Medicine, 31*(3), 398–404. https://doi.org/10.1016/s0891-5849(01)00598-6
25. Konovalova, T. A., Dikanov, S. A., Bowman, M. K., & Kispert, L. D. (2001). Detection of anisotropic hyperfine components of chemically prepared carotenoid radical cations: 1D and 2D ESEEM and pulsed ENDOR study. *The Journal of Physical Chemistry B, 105*, 8361–8358. https://doi.org/10.1021/jp010960n
26. Konovalova, T. A., Gao, Y., Schad, R., Kispert, L. D., Saylor, C. A., & Brunel, L.-C. (2001). Photooxidation of carotenoids in mesoporous MCM-41, Ni-MCM-41 and Al-MCM-41 molecular sieves. *The Journal of Physical Chemistry B, 105*, 7459–7464. https://doi.org/10.1021/jp0108519
27. Gao, Y., Konovalova, T. A., Xu, T., & Kispert, L. D. (2002). Electron transfer of carotenoids imbedded in MCM-41 and Ti-MCM-41: EPR, ENDOR and UV/Vis studies. *The Journal of Physical Chemistry B, 106*, 10808–10815. https://doi.org/10.1021/jp025978s
28. Konovalova, T. A., Gao, Y., Kispert, L. D., van Tol, J., & Brunel, L.-C. (2003). Characterization of Fe-MCM-41 molecular sieves with incorporated carotenoids by multifrequency electron paramagnetic resonance. *The Journal of Physical Chemistry B, 107*, 1006–1011. https://doi.org/10.1021/jp021565f
29. Gao, Y., Konovalova, T. A., Lawrence, J. N., Smitha, M. A., Nunley, J., Schad, R., & Kispert, L. D. (2003). Interactions of carotenoids and Cu^{2+} in Cu-MCM-41: Distance-dependent reversible electron transfer. *The Journal of Physical Chemistry B, 107*, 2459–2465. https://doi.org/10.1021/jp027164h
30. Kispert, L. D., Konovalova, T., & Gao, Y. (2004). Carotenoid radical cations and dications: EPR, optical, and electrochemical studies. *Archives of Biochemistry and Biophysics, 430*(1), 49–60. https://doi.org/10.1016/j.abb.2004.03.036

31. Gao, Y., Kispert, L. D., Konovalova, T. A., & Lawrence, J. N. (2004). Isomerization of carotenoids in the presence of MCM-41 molecular sieves: EPR and HPLC studies. *The Journal of Physical Chemistry B, 108*, 9456–9462. https://doi.org/10.1021/jp036091e
32. Gao, Y., & Kispert, L. D. (2004). Oxidation of carotenoids in MCM-41 and metal ion substituted MCM-41 molecular sieves. *Recent Research Development in Physical Chemistry, 7*, 63–78.
33. Polyakov, N. E., Leshina, T. V., Konovalova, T. A., Hand, E. O., & Kispert, L. D. (2004). Inclusion complexes of carotenoids with cyclodextrins: 1H NMR, EPR, and optical studies. *Free Radical Biology and Medicine, 36*(7), 872–880. https://doi.org/10.1016/j.freeradbiomed.2003.12.009
34. Gao, Y., Kispert, L. D., van Tol, J., & Brunel, L. C. (2005). Electron spin-echo envelope modulation and pulse electron nuclear double resonance studies of Cu^{2+}···beta-carotene interactions in Cu-MCM-41 molecular sieves. *The Journal of Physical Chemistry B, 109*(39), 18289–18292. https://doi.org/10.1021/jp052550v
35. Polyakov, N. E., Leshina, T. V., Salakhutdinov, N. F., Konovalova, T. A., & Kispert, L. D. (2006). Antioxidant and redox properties of supramolecular complexes of carotenoids with beta-glycyrrhizic acid. *Free Radical Biology and Medicine, 40*(10), 1804–1809. https://doi.org/10.1016/j.freeradbiomed.2006.01.015
36. Gao, Y., Focsan, A. L., Kispert, L. D., & Dixon, D. A. (2006). Density functional theory study of the beta-carotene radical cation and deprotonated radicals. *The Journal of Physical Chemistry B, 110*(48), 24750–24756. https://doi.org/10.1021/jp0643707
37. Kispert L. D., & Piekara-Sady, L. (2006). ENDOR Spectroscopy. In: D. R. Vij (Ed.), *Handbook of Applied Solid State Spectroscopy.* Springer.
38. Focsan, A. L., Bowman, M. K., Konovalova, T. A., Molnár, P., Deli, J., Dixon, D. A., & Kispert, L. D. (2008). Pulsed EPR and DFT characterization of radicals produced by photo-oxidation of zeaxanthin and violaxanthin on silica-alumina. *The Journal of Physical Chemistry B, 112*(6), 1806–1819. https://doi.org/10.1021/jp0765650

39. Lawrence, J., Focsan, A. L., Konovalova, T. A., Molnar, P., Deli, J., Bowman, M. K., & Kispert, L. D. (2008). Pulsed electron nuclear double resonance studies of carotenoid oxidation in Cu(II)-substituted MCM-41 molecular sieves. *The Journal of Physical Chemistry B, 112*(17), 5449–5457. https://doi.org/10.1021/jp711310u
40. Polyakov, N. E., Leshina, T. V., Meteleva, E. S., Dushkin, A. V., Konovalova, T. A., & Kispert, L. D. (2009). Water soluble complexes of carotenoids with arabinogalactan. *The Journal of Physical Chemistry B, 113*(1), 275–282. https://doi.org/10.1021/jp805531q
41. Focsan, A. L., Molnár, P., Deli, J., & Kispert, L. (2009). Structure and properties of 9′-*cis* neoxanthin carotenoid radicals by electron paramagnetic resonance measurements and density functional theory calculations: Present in LHC II? *The Journal of Physical Chemistry B, 113*(17), 6087–6096. https://doi.org/10.1021/jp810604s
42. Konovalova, T. A., Li, S., Polyakov, N. E., Focsan, A. L., Dixon, D. A., & Kispert, L. D. (2009). Measuring Ti(III)-carotenoid radical interspin distances in TiMCM-41 by pulsed EPR relaxation enhancement method. *The Journal of Physical Chemistry B, 113*(25), 8704–8716. https://doi.org/10.1021/jp811369h
43. Gao, Y., Shinopoulos, K. E., Tracewell, C. A., Focsan, A. L., Brudvig, G. W., & Kispert, L. D. (2009). Formation of carotenoid neutral radicals in photosystem II. *The Journal of Physical Chemistry B, 113*(29), 9901–9908. https://doi.org/10.1021/jp8075832
44. Polyakov, N. E., Leshina, T. V., Meteleva, E. S., Dushkin, A. V., Konovalova, T. A., & Kispert, L. D. (2010). Enhancement of the photocatalytic activity of TiO_2 nanoparticles by water-soluble complexes of carotenoids. *The Journal of Physical Chemistry B, 114*(45), 14200–14204. https://doi.org/10.1021/jp908578j
45. Kispert, L. D., & Polyakov, N. E. (2010). Carotenoid radicals: Cryptochemistry of natural colorants. *Chemistry Letters, 39*, 148–155. https://doi.org/10.1246/cl.2010.148
46. Polyakov, N. E., Focsan, A. L., Bowman, M. K., & Kispert, L. D. (2010). Free radical formation in novel carotenoid metal ion complexes of astaxanthin. *The Journal of Physical Chemistry B, 114*(50), 16968–16977. https://doi.org/10.1021/jp109039v

47. Kispert, L. D., Focsan, L., & Konovalova, T. (2010). Applications of EPR spectroscopy to understanding carotenoid radicals. In: J. T. Landrum (Ed.), *Carotenoids, Physical Chemical and Biological Functions and Properties.* CRC Press.
48. Focsan, A. L., Bowman, M. K., Molnár, P., Deli, J., & Kispert, L. D. (2011). Carotenoid radical formation: Dependence on conjugation length. *The Journal of Physical Chemistry B, 115*(30), 9495–9506. https://doi.org/10.1021/jp204787b
49. Focsan, A. L., Bowman, M. K., Shamshina, J., Krzyaniak, M. D., Magyar, A., Polyakov, N. E., & Kispert, L. D. (2012). EPR study of the astaxanthin n-octanoic acid monoester and diester radicals on silica-alumina. *The Journal of Physical Chemistry B, 116*(44), 13200–13210. https://doi.org/10.1021/jp307421e
50. Polyakov, N. E., Magyar, A., & Kispert, L. D. (2013). Photochemical and optical properties of water-soluble xanthophyll antioxidants: Aggregation vs complexation. *The Journal of Physical Chemistry B, 117*(35), 10173–10182. https://doi.org/10.1021/jp4062708
51. Focsan, A. L., Polyakov, N. E., & Kispert, L. D. (2021). Carotenoids: Importance in daily life-Insight gained from EPR and ENDOR. *Applied Magnetic Resonance, 52*(8), 1093–1112. https://doi.org/10.1007/s00723-021-01311-8
52. Gao, G., Deng, Y., & Kispert, L. D. (1997). Photoactivated ferric chloride oxidation of carotenoids by near-UV to visible light. *The Journal of Physical Chemistry B, 101*(39), 7844–7849.
53. Gao, Y., & Kispert, L. D. (2003). Reaction of carotenoids and ferric chloride: Equilibria, isomerization, and products. *The Journal of Physical Chemistry B, 107*, 22, 5333–5338. https://doi.org/10.1021/jp034063q
54. Kevan, L., & Kispert, L. D. (1976). *Electron Spin Double Resonance Spectroscopy*, John Wiley & Sons.

5

Carotenoid Complexes

Carotenoids are natural dyes and antioxidants widely used in food processing and in therapeutic formulations. However, practical application is restricted by their high sensitivity to external factors such as heat, light, oxygen, metal ions and processing conditions, as well as their extremely low water solubility. Various approaches have been developed to overcome these problems. In particular, it was demonstrated that application of supramolecular complexes of "host-guest" type with water-soluble nanoparticles allows minimizing the above-mentioned disadvantages. From this point of view, nanoencapsulation of carotenoids is an effective strategy to improve their stability during storage and food processing. Also, nanoencapsulation enhances bioavailability of carotenoids via modulating their release kinetics from the delivery system, influencing their solubility and absorption.

Encapsulation of lipophilic carotenoids with different delivery systems, already performed in numerous studies, is an innovative approach shown to increase their solubility, stability and bioavailability. In a recently published review,[1] we presented the state of the art of carotenoid nanoencapsulation and summarized the data obtained during the past five years on preparation, analysis and reactivity of carotenoids encapsulated into various nanoparticles. When carotenoids are bound to other molecules, they form complexes which can provide unique properties and behaviors. Different delivery systems such as inclusion complexes, or nanoemulsions, nanoliposomes, and biopolymeric nanoparticles have been tested to improve carotenoid properties. The possible mechanisms of carotenoid bioavailability enhancement by multifunctional delivery systems were also discussed.[1] Importantly, some physicochemical studies provided evidences of effect of encapsulation on fundamental properties of carotenoids, redox potentials, quantum yields and lifetime of excited states, optical properties (absorption and fluorescence spectra), as well as properties of paramagnetic forms of carotenoids.

This chapter covers basic concepts of inclusion of carotenoids in "host-guest" complexes such as saccharides. Molecules of oligosaccharides such as cyclodextrins and glycyrrhizic acid, and polysaccharides such as arabinogalactan are capable of forming "host-guest" complexes with carotenoids substantially enhancing their stability, solubility and bioavailability. Such carotenoid complexes studied in collaboration with the University of Alabama, Tuscaloosa[2-11] are described here and discussed in conjunction with other studies that help elucidate encapsulation of carotenoids in these "host-guest" complexes.

Along with enhanced solubility and bioavailability, additional advantages like improvements in chemical stability, protection of carotenoids from external environment, taste modification and controlled release make them interesting delivery systems for further study to incorporate them into beverages, foods, pharmaceuticals or cosmetics.

5.1 Carotenoid Complexes with Cyclodextrins

Cyclodextrins (CDs), a family of cyclic molecules with different degrees of polymerization, including α-CD, β-CD and γ-CD, are the most popular and developed drug delivery systems. They are cyclic oligosaccharides derived from starch with a hydrophilic outer surface and hydrophobic inner cavity suitable for the inclusion of lipophilic "guest" molecules of appropriate size (Figure 5.1). They are widely applied in the pharmaceutical and food industries to enhance the aqueous solubility and dissolution rate of poorly soluble bioactive molecules, as well as for

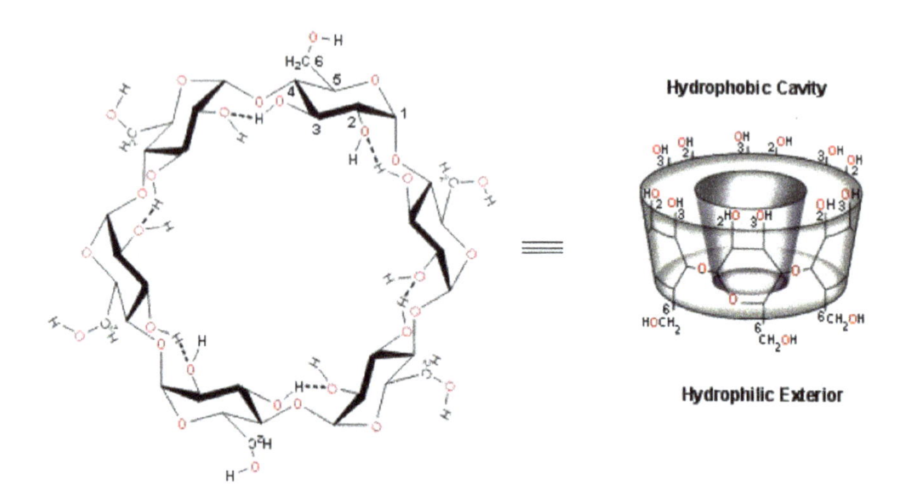

Figure 5.1. Cyclic oligosaccharide cyclodextrin forming a torus-shaped structure with rigid lipophilic cavity in which carotenoids can be included. Adapted with permission from Figure 2 of Reference 1.

increasing their stability and bioavailability. For example, one important application of carotenoids in CDs is in the food and beverage industry. Carotenoids are widely used as natural colorants in food and beverage products, but their instability and poor solubility can pose a problem for manufacturers. The inclusion complexation of carotenoids with CDs can improve their stability and solubility, which allows for more consistent and reliable use as natural colorants. Another application is in the area of nutraceuticals and dietary supplements. Carotenoids are important nutrients that are essential for human health, but they can be difficult to deliver in an effective form. The inclusion complexation of carotenoids with CDs can improve their bioactivity by increasing their stability, solubility, and bioavailability. The inclusion complexation of carotenoids with CDs can also reduce light- and heat-induced degradation and oxidation. CDs are inexpensive, non-toxic, and possess low hygroscopicity and high thermal stability.

The most common CDs are α-CD, β-CD and γ-CD, which have different numbers of glucopyranose units — 6, 7 and 8 units — and different ring diameters of about 6, 8 and 10 Å, respectively. Different CDs have different solubility in water, and their stability and controlled release can be changed by altering their structure, for example by changing the ring size or adding groups.[12] For instance, the water solubility increases significantly by chemical modification of the poorly soluble β-CD to produce hydroxypropyl- or methyl-substituted β-CD.[13] Enhanced storage stability was also shown in earlier studies that involved the carotenoid/CD inclusion complexes. The complex of astaxanthin with hydroxypropyl-β-CD showed storage stability and antioxidant activity but poor solubility.[14]

The encapsulated hydrophobic carotenoids substitute the enthalpy-rich water molecules within the cavity without cleavage or formation of covalent bonds, and they remain in the hydrophobic interior via hydrophobic forces, van der Waals interactions or hydrogen bonds. Non-covalent associations of *trans*-β-carotene complexes with β-CD and γ-CD in water showed evidence for the formation of large aggregates by light scattering and NMR spectroscopy.[15] A study on the lycopene/α- and β-CD complexes by light scattering, ion spray ionization and tandem mass spectrometry also pointed out that large aggregates of particles, on the nanometer-size scale, were present in water, with meaningful differences in the shape when comparing the α-CD with β-CD aggregates.[16] ^{1}H-NMR, EPR and optical studies on inclusion complexes of carotenoids with CDs showed that CD protects the carotenoid from reactive oxygen species but complexation with CD results in considerable decrease in antioxidant ability of the carotenoid.[3] In reality, these complexes form water dispersions, rather than solutions. The reduced color intensity significantly decreases the use of carotenoid/CD complexes, in particular, as food colorants.

Nevertheless, some positive results have been obtained during the last years. In particular, a recent study of carotenoids/CD complexes shows that molecular inclusion of yellow bell pepper carotenoids (lutein, zeaxanthin, α-cryptoxanthin, α-carotene and β-carotene) in the CD cavity provides good results for color protection for beverages compared with the use of the crude extract.[17] The authors have measured the color stability during storage under irradiance and in the absence of light at temperatures of 25–31°C for 21 days and demonstrated higher stability of the color indices in isotonic drinks colored with the inclusion complexes compared to those colored with the crude yellow bell pepper extract. Several carotenoid/β-CD inclusion complexes provided evidence that β-CD inclusion renders carotenoids more stability towards oxidizing agents like 2,2′-azobis(2-methylpropionamidine) dihydrochloride and hydrogen peroxide.[18] Pinzón-García and coauthors[19] investigated the effects of oral administration of the antioxidant carotenoid bixin and its β-CD complex in an obese murine model. The authors conclude that the oral administration of the bixin/β-CD inclusion compound improved the metabolic parameters to be more palatable and hepatoprotective. In another study, Nalawade and Gajjar[20] showed an improved dissolution rate of carotenoid astaxanthin by its complexation with methyl-β-CD using spray drying technique. Application of various physicochemical techniques, FTIR, UV, DSC, ^{1}H-NMR, XRD and molecular modeling analysis confirmed that the side rings of astaxanthin molecule were incorporated into the CD cavity. This study provides the basis for the development of soluble and bioavailable oral formulations of carotenoid complexes with CDs using spray drying technique, which is scalable and accepted in the industry.

5.2 Carotenoid Complexes with Glycyrrhizic Acid

Glycyrrhizic acid (GA) is a saponin extracted from the root of licorice. Like most saponins, the GA molecule has an amphiphilic structure. It contains a hydrophilic part, a glucuronic acid residue, and a hydrophobic part, a glycyrrhetic acid residue (Figure 5.2). GA forms a dimer in a donut-like shape. The polyene chains of carotenoids can reside within the hydrophobic area while allowing the hydrophilic terminal rings to stick out on each end.

Licorice has been used in traditional medicine since ancient times in China, Egypt and Japan for a variety of purposes, including as an anti-inflammatory and anti-viral agent. Nowadays, the biological and therapeutic activity of GA is still intensively investigated, and it was confirmed to be non-toxic for application in medicine and food.[21–23]

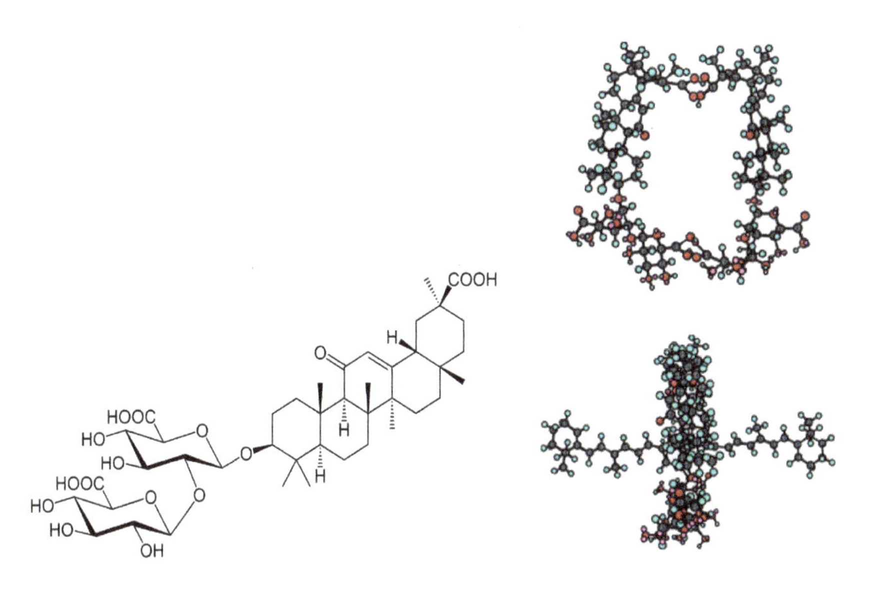

Figure 5.2. The structure of glycyrrhizic acid monomer (left). Suggested structure of the GA dimer and their inclusion complex with the carotenoid (right). Adapted with permission from Figure 3 of Reference 1.

A novel, unusual property of GA has been discovered, namely, its ability to form stable self-associates[24,25] which are able to form water-soluble inclusion complexes with other lipophilic molecules. In particular, it was found that GA is able to form water-soluble supramolecular complexes with a variety of lipophilic drugs.[26] These complexes show significant advantages over other known drug delivery systems, possessing increased solubility, stability and bioavailability. In addition, the membrane-modifying ability of GA has been described.[27–29] The key factor in its therapeutic activity is its ability to incorporate into the lipid bilayer and to increase the membrane fluidity and permeability.[26]

Supramolecular complexes of β-carotene, canthaxanthin, lutein, zeaxanthin, astaxanthin and other carotenoids with GA have been prepared, and their properties were studied by various physicochemical methods (NMR, EPR, electrochemistry, UV spectroscopy).[4,5,9–11,30,31] In these studies, carotenoid-GA complexes were prepared by two different methods. The first method was the traditional "liquid state" technique. Since GA is soluble both in water and organic solvents, a solution of carotenoid in methylene chloride was added to the aqueous or organic GA solution and then stirred for several hours for complex preparation. The second "solid state" method of complex preparation was a mechanochemical treatment of the solid mixture of carotenoid crystals with GA crystals. Co-grinding of powder components results in formation of solid dispersion without use of any organic solvents. In such processes, hydrogen bonds, π-stacking, van der Waals, ion pairing interactions, etc., are broken, leading to formation of supramolecular complexes directly in solid state or hybrid molecular crystals.

Carotenoids are able to form inclusion complexes with GA micelles at high concentrations (>1 mM of GA), as well as with pre-micellar GA aggregates (dimers) at low concentrations (1 μM–1 mM).[4] These complexes are extremely stable. The stability constants of carotenoid-GA complexes are 1–2 orders higher than stability constants of carotenoid-CD complexes.[4]

Encapsulation of carotenoids lutein and zeaxanthin into GA micelles protects them from oxidation by reactive oxygen species like ozone and hydroxyl radicals, and metal ions.[30] For example, oxidation rate of lutein and zeaxanthin by ozone molecules in aqueous-ethanol solution decreased 10 times in the presence of 1 mM of GA. Similar effects were detected in the presence of disodium salt of GA: in the presence of 1 mM of Na_2GA the oxidation rate of lutein and zeaxanthin by Fe^{3+} ions decreased by 10–20 times.[30]

In contrast to most of the water-soluble oligosaccharides and polysaccharides,[10,11] GA is able to form supramolecular complexes with carotenoids not only in aqueous solutions where GA complexes increased the carotenoid solubility more than 1,000-fold,[30] but also in non-aqueous organic solvents (alcohols, DMSO, acetonitrile).[4,10,11] This fact is important for discussing the possibility of GA-assisted transport of carotenoid molecules through lipophilic cell membranes and their membrane protection properties.

In organic solvents GA is able to form supramolecular complexes not only with neutral carotenoid molecules, but also with their paramagnetic forms — radical cations and charge transfer complexes with electron donors.[4] Also, there is a significant increase in the lifetime of β-carotene radical cations (50-fold) in the presence of GA. High stability of the carotenoid radical cations imbedded into GA host opens up possibilities for the application of these complexes for the design of artificial light-harvesting, photoredox and catalytic systems.

One of the most important biological properties of carotenoids is their antioxidant activity. In aqueous environment as well as in lipid membranes, carotenoids trap toxic oxygen radicals and thus prevent damage to the living organism. Some studies were performed to elucidate how the complexation with GA affects the ability of carotenoids to scavenge reactive oxygen radicals.[5,9] The antioxidant activity of carotenoid complexes was studied by the EPR spin-trapping technique. The details of this technique and effectiveness in the measurement of scavenging rates towards hydroperoxyl OOH radical were described in our earlier study.[2]

Comparison of the scavenging rates of hydroperoxyl radicals by free carotenoids and their GA complexes in non-aqueous solution (DMSO) shows a strong dependence of the rate constants on the carotenoid structures and their oxidation potentials (Table 5.1).

In a recent study the oxidation potential of *cis*-bixin referenced to ferrocene (0.94 V) was listed as the highest measured oxidation potential of all carotenoids studied.[32] Using this oxidation potential and extrapolating the scavenging ability plot predicts a relative scavenging value for *cis*-bixin of 44, or 501 assuming a semi-log plot. This sets bixin as the naturally occurring carotenoid with the highest scavenging rate, even higher than that of the synthesized 7′-apo-7′,7′-dicyano-β-carotene. 7′-apo-7′,7′-dicyano-β-carotene was found to be very stable when tuning lasers in Fleming's lab at the University of California, Berkeley. Similarly, we expect bixin to be very stable. Along with its highest scavenging ability

Table 5.1. Relative scavenging rate constants of OOH radicals by carotenoids and their GA complexes (k/k_{ST}) in DMSO. $E_{1/2}$ is the redox potential of carotenoids (in V vs. SCE) referenced to ferrocene.[5,9]

[GA], mM	β-carotene	Canthaxanthin	7′-apo-7′,7′-dicyano- β-carotene
	($E_{1/2}$ = 0.634 V)	($E_{1/2}$ = 0.765 V)	($E_{1/2}$ = 0.825 V)
0	0.5	2	7
0.5	0.5	59	133

and its potential to increase shelf life, bixin makes a great candidate as a dye in foods, drugs or cosmetics. Furthermore, very little amounts are needed when used as coloring agent because *cis*-bixin has a high extinction coefficient of 10^5.[33]

Since it was found[5] that scavenging ability of carotenoids towards hydroperoxyl radicals is strongly potential dependent, it was suggested that GA complexation can affect the oxidation potential of the carotenoids. This hypothesis was proved by the measurement of the oxidation potential of two carotenoids zeaxanthin and canthaxanthin in the presence of GA. In both cases, an increase in $E_{1/2}$ by 0.03–0.05 V was observed.[5] It was assumed that interaction between carotenoids and hydroperoxyl radicals occurs via hydrogen abstraction from the most acidic proton at the C4 position of carotenoids.[9,34] GA forms a donut-like dimer in which the hydrophobic polyene chain of carotenoids lies protected within the donut hole, permitting the hydrophilic ends and most acidic proton to be exposed to the surroundings. Note that in the case of the CD complexes, the terminal group of the carotenoid is completely protected which results in the inhibition of antioxidant activity.[3]

The characteristic feature of xanthophyll carotenoids (lutein, zeaxanthin, astaxanthin, β-cryptoxanthin) is their ability to self-assemble into J- and H-type aggregates in aqueous solution and in lipid membrane.[9] The formation of such aggregates occurs even in organic solvents in the presence of small amounts of water (5–10%) and significantly changes the photophysical and optical properties of these carotenoids which are important for solar energy conversion and light-induced oxidative damage. As an example, in the absorption spectrum of zeaxanthin, a large shift of absorption band from 460 nm to 380 nm with the loss of vibrational structure of the S2 excited state indicates the presence of the zeaxanthin dimers. In contrast, the optical study of *cis*-bixin[35] shows minimal problem with aggregation in mixtures contrary to some *trans* carotenoids like zeaxanthin. Due to the unfavorable *cis* ground state structure, there is a lack of H-aggregation but there is a 4 nm red shift in a 50% water-methanol solution, suggesting J-aggregation. The optical shifts with solvent depend on the polarity of the solvent. Lack of H-aggregation prevents a large blue shift, which would allow for more blue light absorbance, possibly causing decomposition. Also, the antioxidant activity of xanthophyll carotenoids is significantly reduced due to aggregation reaction.[9] This feature might be important for antioxidant and photoprotective function of these carotenoids in lipid membranes. In particular, it was shown by the EPR technique that in the presence of Fe^{2+} ions and hydrogen peroxide (Fenton reaction), zeaxanthin aggregates can demonstrate pro-oxidant effects instead of antioxidant activity.[9] It was demonstrated by UV spectroscopy that complexation with GA reduces the aggregation ability of the xanthophyll carotenoids.[9] Simultaneously, the scavenging ability of xanthophyll carotenoids to trap free oxygen radicals increases since GA prevents aggregation, allowing the cyclohexene ring of carotenoids to be exposed to the surroundings. Considering the important role of xanthophylls in eye and skin health, GA should be considered as a perspective delivery system to provide enhanced solubility and activity of the carotenoids.

Recent studies have shown that GA is able not only to form inclusion complexes with carotenoids and various drug molecules, but also to modify the structure of lipid membranes.[28–30] Using NMR relaxation techniques and molecular dynamics simulation it was demonstrated that GA molecules are able to penetrate into the lipid bilayer and affect the lipid dynamics. Taking into account the protective role of some carotenoids in lipid membranes and the ability of GA to form stable associates with carotenoids in non-aqueous environments, one can expect the influence of GA on the antioxidant ability of carotenoids in membranes. In this case the ability of GA to prevent the xanthophyll carotenoids aggregation and thus to enhance their antioxidant properties[9] also can play a certain positive role.

5.3 Carotenoid Complexes with Arabinogalactan

Another "host molecule" for carotenoid encapsulation and delivery that was studied is polysaccharide arabinogalactan (AG). AG is a natural branched polymer (Figure 5.3) consisting of arabinose and galactose units with a molecular mass near 16 kDa extracted primarily from larch. It has been used in traditional medicine for a variety of purposes, including as

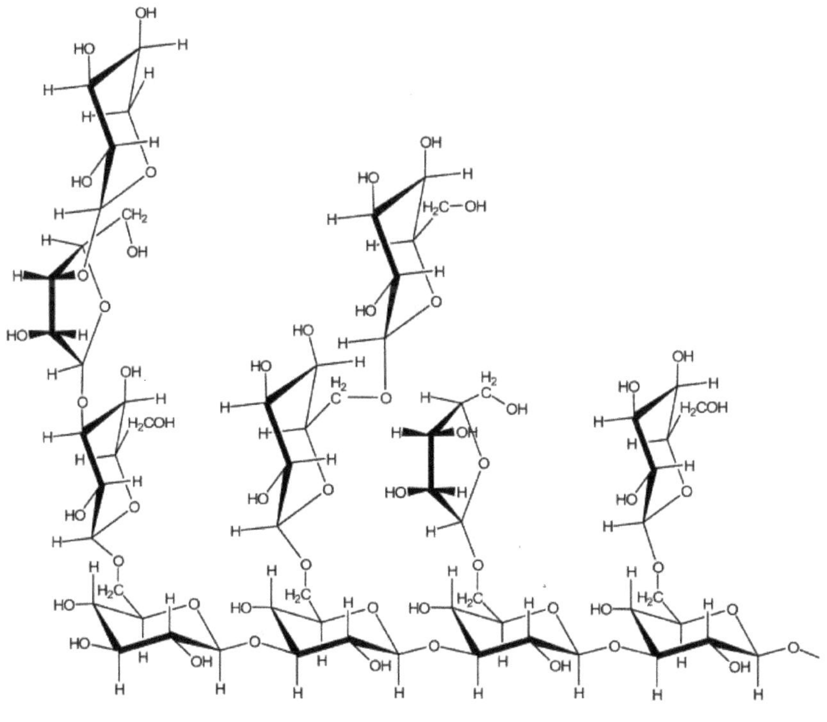

Figure 5.3. Fragment of a branched polysaccharide arabinogalactan. Adapted with permission from Figure 4 of Reference 1.

an immune system booster and a prebiotic. There are clinical results that suggest a role for larch AG in the improvement of immune system and defense against pathogens in humans.[36]

Larch AG is non-toxic and approved by the U.S. Food and Drug Administration as an important source of dietary fiber. AG is biodegradable, biocompatible, hydrophilic, and contains different functional groups such as hydroxyl and carboxylic acid that make it ideal for conjugation and delivery of carotenoids and drug molecules. There are many examples of increased solubility of hydrophobic drugs in the complexes with AG.[37–39] Some recent studies indicate also that AG has good adhesion to membrane surfaces, and is able to enhance the permeability for both human and plant cell membranes.[27,40]

AG is highly water soluble and produces low-viscosity solutions. Since AG is soluble only in water, but carotenoids are insoluble in water, the solid-state mechanochemical technique had been developed for preparation of carotenoid-AG inclusion complexes directly in the solid state without use of any organic solvents.[6,8,9,30,31] Co-grinding of the solid polysaccharide with carotenoids results in penetration of the carotenoid molecules into the "host" polysaccharide, and formation of a water-soluble carotenoid complex. As a result, the solubility of β-carotene and canthaxanthin complexes with AG prepared for the first time[6] were 2–5 mM in aqueous solution, which is six orders of magnitude higher than the characteristic solubility of free carotenoids in pure water (~1 nM). The mechanochemical method for solid-state complex preparation has evident advantages compared with traditional "liquid phase" techniques. The interest in solvent-free conditions stems from the possibility of producing the same product as that from solution without solvent because this process is cheaper, less time consuming and more environmentally friendly. In the case of carotenoids, the solid-state technique opens up the possibility of obtaining products not otherwise accessible in solutions.

As was mentioned, one of the main problems with the practical application of carotenoids is their chemical instability in the presence of light, oxygen, water, and metal ions. Although they are considered as effective photoprotectors of living cells, in aerated aqueous solution they are unstable and would be unacceptable as colorants and antioxidants in foods. Various physicochemical techniques have been applied to study the reactivity of carotenoids incorporated into AG macromolecule in aqueous solution. The carotenoid-AG complexes maintain their original color and show insignificant changes in absorption spectra.[6,9] It was demonstrated also that complexation with AG prevents H-aggregate formation of xanthophyll carotenoids in the presence of water as was detected for carotenoid complexes with GA.[9] It was found that AG complexes have enhanced photostability and oxidation stability compared to pure carotenoids. We suggest that these results are important for a variety of carotenoid applications. A significant increase (5–10 times) in photostability of carotenoids lutein, astaxanthin and canthaxanthin in the inclusion complex with AG was detected.[6,9]

As an example, Figure 5.4 demonstrates the difference in the photodegradation rate of astaxanthin in pure form and in the complex with AG ([AG] = 0.1 mM) in the presence and absence of irradiation. The authors proposed that the main mechanism of enhanced photostability of carotenoids in a polysaccharide complex is their isolation from water by incorporation into the hydrophobic polymer environment.

Incorporation of carotenoids into the hydrophobic polymer environment results also in enhancement of their chemical stability. The complete inhibition of oxidation of the carotenoids lutein and zeaxanthin by Fe^{3+} ions as an electron acceptor and by ozone molecules as powerful oxidant

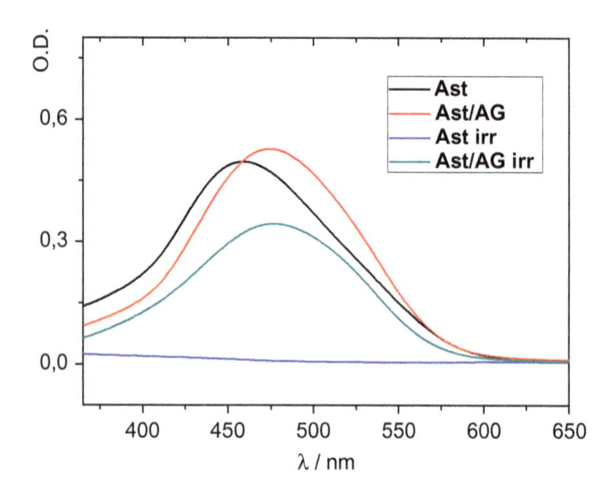

Figure 5.4. Photodegradation of astaxanthin in aerated 25% ethanol solution irradiated for 1 minute by the full light of a xenon lamp. Adapted with permission from Reference 9.

was achieved by complexation with AG.[30] Oxidation of carotenoids by Fe^{3+} ions results in formation of carotenoid radical cations, and this reaction was described in details in our earlier studies.[4,41] On the other hand, radical cations of carotenoids are unstable in aqueous solutions due to fast deprotonation and formation of carotenoid neutral radicals.[42] Figure 5.5 demonstrates an example of the difference in the oxidation rate of lutein by ozone in pure form and in the complex with AG in aqueous solution.

Ozone molecules are known to react with unsaturated double bonds with high efficiency.[43,44] The products of this reaction are unstable ozonides which decay to final oxidation products — ketones, aldehydes, and carboxylic acids. As an example, Scheme 5.1 shows the reaction pathway in aqueous solution.

The authors[30] suggested two main factors responsible for chemical stabilization of carotenoids in the AG complexes, namely, isolation from water and isolation from reactive species (ozone molecules and metal

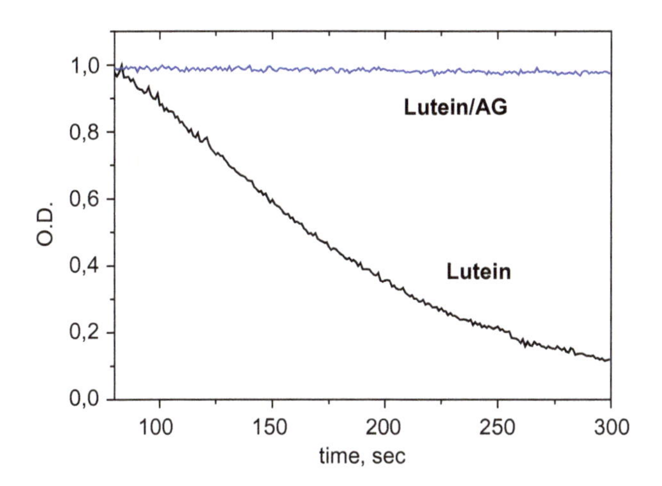

Figure 5.5. Oxidation of lutein and lutein/AG complex by ozone in 25% ethanol solution. Concentration of AG 0.05 mM. Adapted with permission from Reference 30.

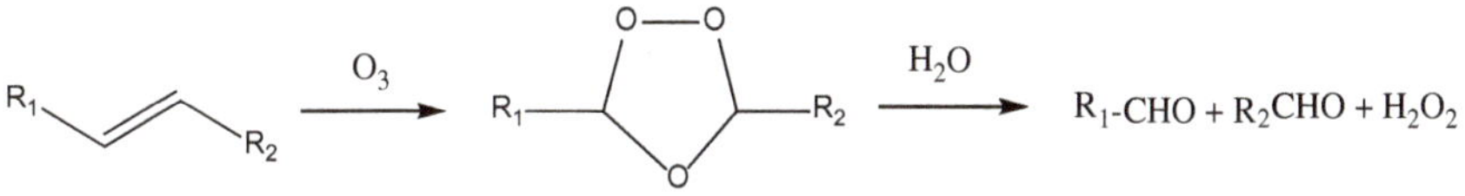

R_1–CH=CH–R_2 $\xrightarrow{O_3}$ R_1–(ozonide: O–O, O ring)–R_2 $\xrightarrow{H_2O}$ R_1-CHO + R_2CHO + H_2O_2

Scheme 5.1. The mechanism of decay for the conjugated double bonds of carotenoids and other unsaturated compounds in the presence of ozone in water solution. Adapted with permission from Reference 44.

ions). We assume that the stability of the carotenoids incorporated into the polysaccharide macromolecule might have wide practical applications.

An important question is whether the described features of the complexes (solubility, membrane permeability, stability) affect the bioavailability of carotenoids in living systems. To answer this question, Li and coauthors[31] investigated the retinal accumulation of macular carotenoids in wild-type mice when fed with zeaxanthin and lutein using an optimized carotenoid-feeding method. Macular carotenoids are well-known natural antioxidants and light-screening compounds that are capable of

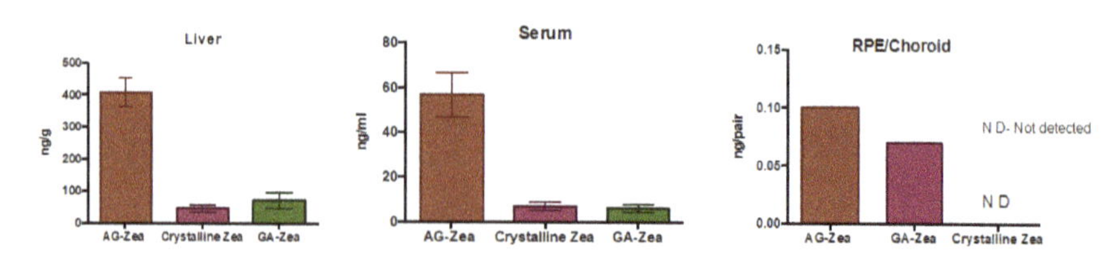

Figure 5.6. Comparison of three carotenoid delivery methods in C57BL/6 wild-type mice. 12-week-old wild-type mice were fed with chow containing zeaxanthin in the formulations of crystalline carotenoid, complex with GA (1:20) and complex with AG (1:20). Values indicate means ± standard deviation from 5 to 10 mice in each group. RPE is the pigmented cell layer just outside the neurosensory retina that nourishes retinal visual cells, and is firmly attached to the underlying choroid and overlying retinal visual cells. Adapted with permission from Reference 31.

quenching singlet oxygen and other free radicals and absorbing potentially damaging blue light.[45] Supplementation with macular carotenoids can prevent and reduce the risk of age-related macular degeneration and other ocular diseases. High-performance liquid chromatography analysis revealed that zeaxanthin and lutein were detected in serum, liver and retinal pigment epithelium (RPE)/choroid of the mice. Significantly higher amounts of lutein and zeaxanthin were detected when they use carotenoid/AG complex instead of crystalline carotenoids (Figure 5.6). Interestingly, only the GA and AG complexes are able to successfully deliver zeaxanthin into the RPE/choroid of wild-type mice.

We can conclude that AG and GA are prospective delivery systems for various carotenoid applications. These complexes demonstrated enhanced solubility and stability in aqueous solutions, as well as enhanced bioavailability in model animal studies. Table 5.2 summarizes studies of carotenoids encapsulated in CD, GA and GA presented here.

Carotenoids are hydrophobic, chemically unstable molecules which are prone to oxidation. The presence of conjugated double bonds in the

Table 5.2. CD, GA and AG delivery systems.

Delivery system	Carotenoid	Encapsulation method	Results	Reference
β-CD	β-caroten-8′-oic acid	Mixture of ethanol solution of carotenoid with aqueous CD solution	CD protects the carotenoid from reactive oxygen species. Complexation results in considerable decrease in antioxidant ability of the carotenoid.	3
AG	β-carotene canthaxanthin and 7′-apo-7′,7′-dicyano-β-carotene	Solid-state mechanochemical method	Enhanced photostability by a factor of 10 in water solution, increased stability by a factor of 20 toward metal ions in solution, increase in yield and stability of the radical cation in AG.	5
GA, AG	astaxanthin, lutein, zeaxanthin	Mixture of ethanol solution of carotenoid with aqueous solution of AG or GA	Solubility enhancement, prevention of H-aggregate formation in ethanol/water mixture, 7-fold increase of photostability in solution.	4 9
GA, Na_2GA and AG	lutein, zeaxanthin	Solid-state mechanochemical method	2,000-fold solubility enhancement, more than 10-fold increase of carotenoid stability in solution toward oxidation by ozone and Fe ions.	30
GA, AG	lutein, zeaxanthin, and β-carotene	Solid-state mechanochemical method	Solubility and bioavailability enhancement, increase of zeaxanthin level in the serum, liver and RPE/choroid of mice.	31

carotenoid structure contributes to their pigmentation, absorption of UV-vis radiation, and antioxidant activity but also is the main reason for their chemical instability. The efficient way to increase the bioavailability of carotenoids is to increase their solubility and stability by encapsulation into water-soluble carriers. Encapsulation refers to a physicochemical process to entrap active compounds in the environment of a compound which can enhance their efficiency, specificity and targeting ability. At this moment there are already a large number of examples to show that encapsulation of lipophilic carotenoids with nanocarriers as delivery systems is an innovative approach to increase their solubility, stability and bioavailability features important for practical applications. Also, encapsulation affects many fundamental physical and chemical properties of carotenoids, including optical properties, self-association ability, oxidation potentials, stability of paramagnetic forms, and many others described in the present chapter. We can summarize that there is no universal multifunctional carrier which is able to improve all carotenoid features simultaneously. Most delivery systems improve carotenoid solubility, especially by incorporation into water-soluble carriers like CD, AG or GA. Some of the carriers improve carotenoid stability during storage and food processing, as well as their bioavailability in *in vivo* experiments (both oral and topical delivery routes).

References

1. Focsan, A. L., Polyakov, N. E., & Kispert, L. D. (2019). Supramolecular carotenoid complexes of enhanced solubility and stability-The way of bioavailability improvement. *Molecules, 24*(21), 3947. https://doi.org/10.3390/molecules24213947
2. Polyakov, N. E., Leshina, T. V., Konovalova, T. A., & Kispert, L. D. (2001). Carotenoids as scavengers of free radicals in a Fenton reaction: Antioxidants or pro-oxidants? *Free Radical Biology and Medicine, 31*(3), 398–404. https://doi.org/10.1016/s0891-5849(01)00598-6

3. Polyakov, N. E., Leshina, T. V., Konovalova, T. A., Hand, E. O., & Kispert, L. D. (2004). Inclusion complexes of carotenoids with cyclodextrins: 1H NMR, EPR, and optical studies. *Free Radical Biology and Medicine, 36*(7), 872–880. https://doi.org/10.1016/j.freeradbiomed.2003.12.009
4. Polyakov, N. E., Leshina, T. V., Salakhutdinov, N. F., & Kispert, L. D. (2006). Host-guest complexes of carotenoids with β-glycyrrhizic acid. *The Journal of Physical Chemistry B, 110*(13), 6991–6998. https://doi.org/10.1021/jp056038
5. Polyakov, N. E., Leshina, T. V., Salakhutdinov, N. F., Konovalova, T. A., & Kispert, L. D. (2006). Antioxidant and redox properties of supramolecular complexes of carotenoids with beta-glycyrrhizic acid. *Free Radical Biology and Medicine, 40*(10), 1804–1809. https://doi.org/10.1016/j.freeradbiomed.2006.01.015
6. Polyakov, N. E., Leshina, T. V., Meteleva, E. S., Dushkin, A. V., Konovalova, T. A., & Kispert, L. D. (2009). Water soluble complexes of carotenoids with arabinogalactan. *The Journal of Physical Chemistry B, 113*(1), 275–282. https://doi.org/10.1021/jp805531q
7. Kispert, L. D., & Polyakov N. E. (2010). Carotenoid radicals: Cryptochemistry of natural colorants. *Chemistry Letters, 39*(3), 148–155. https://doi.org/10.1246/cl.2010.148
8. Polyakov, N. E., Leshina, T. V., Meteleva, E. S., Dushkin, A. V., Konovalova, T. A., & Kispert, L. D. (2010). Enhancement of the photocatalytic activity of TiO_2 nanoparticles by water-soluble complexes of carotenoids. *The Journal of Physical Chemistry B, 114*(45), 14200–14204. https://doi.org/10.1021/jp908578j
9. Polyakov, N. E., Magyar, A., & Kispert, L. D. (2013). Photochemical and optical properties of water-soluble xanthophyll antioxidants: Aggregation vs complexation. *The Journal of Physical Chemistry B, 117*(35), 10173–10182. https://doi.org/10.1021/jp4062708
10. Polyakov, N. E., & Kispert, L. D. (2015). Water soluble biocompatible vesicles based on polysaccharides and oligosaccharides inclusion complexes for carotenoid delivery. *Carbohydrate Polymers, 128*, 207–219. https://doi.org/10.1016/j.carbpol.2015.04.016
11. Polyakov, N. E., & Kispert, L. D. (2019). Water Soluble Supramolecular Complexes of β-Carotene and other Carotenoids. In: Haugen, L., & Bjornson,

T. (Eds.), *Beta Carotene: Dietary Sources, Cancer and Cognition.* Nova Science Publishers.

12. Saokham, P., Muankaew, C., Jansook, P., & Loftsson, T. (2018). Solubility of cyclodextrins and drug/cyclodextrin complexes. *Molecules, 23*(5), 1161. https://doi.org/10.3390/molecules23051161
13. Gharibzahedi, S. M. T., & Jafari, S. M. (2017). Nanocapsule Formation by Cyclodextrins. In: Jafari, S. M. (Ed.), Nanoencapsulation Technologies for the Food and Nutraceutical Industries. Academic Press. https://doi.org/10.1016/B978-0-12-809436-5.00007-0
14. Yuan, C., Du, L., Jin, Z., & Xu, X. (2013). Storage stability and antioxidant activity of complex of astaxanthin with hydroxypropyl-β-cyclodextrin. *Carbohydrate Polymers, 91*, 385–389. https://doi.org/10.1016/j.carbpol.2012.08.059
15. Mele, A., Mendichi, R., & Selva, A. (1998). Non-covalent associations of cyclomaltooligosaccharides (cyclodextrins) with trans-β-carotene in water: Evidence for the formation of large aggregates by light scattering and NMR spectroscopy. *Carbohydrate Research, 310*(4), 261–267. https://doi.org/10.1016/S0008-6215(98)00193-1
16. Mele, A., Mendichi, R., Selva, A., Molnar, P., & Toth, G. (2002). Non-covalent associations of cyclomaltooligosaccharides (cyclodextrins) with carotenoids in water. A study on the alpha- and beta-cyclodextrin/psi,psi-carotene (lycopene) systems by light scattering, ionspray ionization and tandem mass spectrometry. *Carbohydrate Research, 337*(12), 1129-1136. https://doi.org/10.1016/s0008-6215(02)00097-6
17. Lobo, F. A. T., Silva, V., Domingues, J., Rodrigues, S., Costa, V., Falcγo, D., & de Lima Araújo, K. G. (2018). Inclusion complexes of yellow bell pepper pigments with β-cyclodextrin: Preparation, characterisation and application as food natural colorant. *Journal of the Science of Food and Agriculture, 98*(7), 2665–2671. https://doi.org/10.1002/jsfa.8760
18. Fernández-García, E., & Pérez-Gálvez, A. (2017). Carotenoid: β-cyclodextrin stability is independent of pigment structure. *Food Chemistry, 221*, 1317–1321. https://doi.org/10.1016/j.foodchem.2016.11.024
19. Pinzón-García, A. D., Orellano, L. A. A., de Lazari, M. G. T., Campos, P. P., Cortes, M. E., & Sinisterra, R. D. (2018). Evidence of hypoglycemic, lipid-lowering and hepatoprotective effects of the Bixin and Bixin: β-CD inclusion

compound in high-fat-fed obese mice. *Biomedicine & Pharmacotherapy, 106*, 363–372. https://doi.org/10.1016/j.biopha.2018.06.144

20. Nalawade, P., & Gajjar, A. (2015). Assessment of *in-vitro* bio accessibility and characterization of spray dried complex of astaxanthin with methylated betacyclodextrin. *Journal of Inclusion Phenomena and Macrocyclic Chemistry, 83*, 63–75. https://doi.org/10.1007/s10847-015-0541-8
21. Shibata, S. (2000). A drug over the millennia: pharmacognosy, chemistry, and pharmacology of licorice. *Journal of the Pharmaceutical Society of Japan, 120*(10), 849–862. https://doi.org/10.1248/yakushi1947.120.10_849
22. Ming, L. J., & Yin, A. C. (2013). Therapeutic effects of glycyrrhizic acid. *Natural Product Communications, 8*(3), 415–418.
23. Su, X., Wu, L., Hu, M., Dong, W., Xu, M., & Zhang, P. (2017). Glycyrrhizic acid: A promising carrier material for anticancer therapy. *Biomedicine & Pharmacotherapy, 95*, 670–678. https://doi.org/10.1016/j.biopha.2017.08.123
24. Petrova, S. S., Schlotgauer, A. A., Kruppa, A. I., & Leshina, T. V. (2016). Self-association of glycyrrhizic acid. NMR Study. *Zeitschrift für Physikalische Chemie, 231*, 1-17. https://doi.org/10.1515/zpch-2016-0845
25. Borisenko, S. N., Lekar', A. V., Vetrova, E. V., & Filonova, O. V. (2016). A mass spectrometry study of the self-association of glycyrrhetinic acid molecules. *Russian Journal of Bioorganic Chemistry, 42*, 716–720. https://doi.org/10.1134/S1068162016070037
26. Selyutina, O. Y., & Polyakov, N. E. (2019). Glycyrrhizic acid as a multifunctional drug carrier — from physicochemical properties to biomedical applications: A modern insight on the ancient drug. *International Journal of Pharmaceutics, 559*, 271–279. https://doi.org/10.1016/j.ijpharm.2019.01.047
27. Selyutina, O. Y., Apanasenko, I. E., Shilov, A. G., Khalikov, S. S., & Polyakov, N. E. (2017). Effect of natural polysaccharides and oligosaccharides on the permeability of cell membranes. *Russian Chemical Bulletin, 66*, 129–135. https://doi.org/10.1007/s11172-017-1710-2
28. Selyutina, O. Y., Apanasenko, I. E., Kim, A. V., Shelepova, E. A., Khalikov, S. S., & Polyakov, N. E. (2016). Spectroscopic and molecular dynamics characterization of glycyrrhizin membrane-modifying activity. *Colloids and Surfaces B: Biointerfaces, 147*, 459–466. https://doi.org/10.1016/j.colsurfb.2016.08.037

29. Selyutina, O. Y., Polyakov, N. E., Korneev, D. V., & Zaitsev, B. N. (2016). Influence of glycyrrhizin on permeability and elasticity of cell membrane: perspectives for drugs delivery. *Drug Delivery, 23*(3), 858–865. https://doi.org/10.3109/10717544.2014.919544
30. Apanasenko, I. E., Selyutina, O. Y., Polyakov, N. E., Suntsova, L. P., Meteleva, E. S., Dushkin, A. V., Vachali, P., & Bernstein, P. S. (2015). Solubilization and stabilization of macular carotenoids by water soluble oligosaccharides and polysaccharides. *Archives of Biochemistry and Biophysics, 572*, 58–65. https://doi.org/10.1016/j.abb.2014.12.010
31. Li, B., Vachali, P. P., Shen, Z., Gorusupudi, A., Nelson, K., Besch, B. M., Bartschi, A., Longo, S., Mattinson, T., Shihab, S., Polyakov, N. E., Suntsova, L. P., Dushkin, A. V., & Bernstein, P. S. (2017). Retinal accumulation of zeaxanthin, lutein, and β-carotene in mice deficient in carotenoid cleavage enzymes. *Experimental Eye Research, 159*, 123–131. https://doi.org/10.1016/j.exer.2017.02.016
32. Tay-Agbozo, S., Street, S., & Kispert, L. (2018). The carotenoid bixin found to exhibit the highest measured carotenoid oxidation potential to date consistent with its practical protective use in cosmetics, drugs and food. *Journal of Photochemistry and Photobiology B: Biology, 186*, 1–8. https://doi.org/10.1016/j.jphotobiol.2018.06.016
33. Tay-Agbozo, S., Street, S. C., & Kispert, L. D. (2018). Diffuse-reflectance infrared Fourier transform and electron nuclear double resonance study of the carotenoid bixin attached to irradiated TiO_2. *The Journal of Physical Chemistry C, 122*(33), 19075–19081. https://doi.org/10.1021/acs.jpcc.8b06240
34. Focsan, A. L., Bowman, M. K., Shamshina, J., Krzyaniak, M. D., Magyar, A., Polyakov, N. E., & Kispert L. D. (2012). EPR study of the astaxanthin n-octanoic acid monoester and diester radicals on silica–alumina. *The Journal of Physical Chemistry, 116*, 13200–13210. https://doi.org/10.1021/jp307421e
35. Tay-Agbozo, S., Street, S., & Kispert, L. D. (2018). The carotenoid bixin: Optical studies of aggregation in polar/water solvents. *Journal of Photochemistry and Photobiology A: Chemistry*, 362, 31–39. https://doi.org/10.1016/j.jphotochem.2018.05.008

36. Dion, C., Chappuis, E., & Ripoll, C. (2016). Does larch arabinogalactan enhance immune function? A review of mechanistic and clinical trials. *Nutrition & Metabolism, 13*, 28. https://doi.org/10.1186/s12986-016-0086-x
37. Dushkin, A. V., Tolstikova, T. G., Khvostov, M. V., & Tolstikov, G. A. (2012). Complexes of Polysaccharides and Glycyrrhizic Acid with Drug Molecules-Mechanochemical Synthesis and Pharmacological Activity. In: Karunaratne, D. N. (Ed.), *The Complex World of Polysaccharides*. IntechOpen, pp. 573–602.
38. Kong, R., Zhu, X., Meteleva, E. S., Polyakov, N. E., Khvostov, M. V., Baev, D. S., Tolstikova, T. G., Dushkin, A. V., & Su, W. (2018). Atorvastatin calcium inclusion complexation with polysaccharide arabinogalactan and saponin disodium glycyrrhizate for increasing of solubility and bioavailability. *Drug Delivery and Translational Research, 8*(5), 1200–1213. https://doi.org/10.1007/s13346-018-0565-x
39. Khvostov, M. V., Borisov, S. A., Tolstikova, T. G., Dushkin, A. V., Tsyrenova, B. D., Chistyachenko, Y. S., Polyakov, N. E., Dultseva, G. G., Onischuk, A. A., & An'kov, S. V. (2017). Supramolecular complex of ibuprofen with larch polysaccharide arabinogalactan: studies on bioavailability and pharmacokinetics. *European Journal of Drug Metabolism and Pharmacokinetics, 42*(3), 431–440. https://doi.org/10.1007/s13318-016-0357-y
40. Selyutina, O. Y., Apanasenko, I. E., Khalikov, S. S., & Polyakov, N. E. (2017). Natural poly- and oligosaccharides as novel delivery systems for plant protection compounds. *Journal of Agricultural and Food Chemistry, 65*, 6582–6587. https://doi.org/10.1021/acs.jafc.7b02591
41. Gao, Y., & Kispert, L. D. (2003). Reaction of carotenoids and ferric chloride: Equilibria, isomerisation, and products. *The Journal of Physical Chemistry B, 107*, 5333–5338. https://doi.org/10.1021/jp034063q
42. Gao, Y., Webb, S., & Kispert, L. D. (2003). Deprotonation of carotenoid radical cation and formation of a didehydrodimer. *The Journal of Physical Chemistry B, 107*, 13237–13240. https://doi.org/10.1021/jp0358679
43. Gluschenko, O. Y., Polyakov, N. E. & Leshina, T. V. (2011). NMR relaxation study of cholesterol binding with plant metabolites. *Applied Magnetic Resonance, 41*, 283–294. https://doi.org/10.1007/s00723-011-0258-9

44. Henry, L. K., Puspitasari-Nienaber, N. L., Jarén-Galαn, M., van Breemen, R. B., Catignani, G. L., & Schwartz, S. J. (2000). Effects of ozone and oxygen on the degradation of carotenoids in an aqueous model system. *Journal of Agricultural and Food Chemistry*, *48*(10), 5008–5013. https://doi.org/10.1021/jf000503o
45. Bernstein, P. S., Li, B., Vachali, P. P., Gorusupudi, A., Shyam, R., Henriksen, B. S., & Nolan, J. M. (2016). Lutein, zeaxanthin, and meso-zeaxanthin: The basic and clinical science underlying carotenoid-based nutritional interventions against ocular disease. *Progress in Retinal and Eye Research*, *50*, 34–66. https://doi.org/10.1016/j.preteyeres.2015.10.003

6

Photoprotection by Carotenoid Radicals

Carotenoids are vital molecules for plant photosynthesis and photoprotection and are essential for survival. The results presented in previous chapters support the formation of the neutral radicals $^{\#}Car^{\bullet}$ of carotenoids *in vitro* on solid surfaces. But do these neutral radicals occur *in vivo*? If they occur *in vivo*, they could provide another effective non-photochemical quencher of the singlet and triplet excited states of chlorophyll, photoprotecting the plant in the presence of excess light. There are several studies[1–10] supporting the formation of neutral radicals from the radical cations and hypothesizing the presence of neutral radicals *in vivo* and their role in photoprotection. Our review articles[7,8,10] summarize all the key concepts learned about radical cations and neutral radicals that are necessary for photoprotection. As confirmed by DFT calculations, proton loss in carotenoids is most favorable at positions situated at the terminal ends of the radical cation.[2,5] This feature was especially useful in analyzing crystal structures of light-harvesting complexes (LHCs) for possible proton loss from a carotenoid radical cation. Figure 6.1 shows the crystal structure of the major LHCII with four carotenoids: zeaxanthin, violaxanthin, lutein and 9′-*cis*-neoxanthin. For proton loss to occur *in vivo*, the orientation of the carotenoids in thylakoid membranes is especially important. Figure 6.1(a) shows the orientation of the carotenoids in LHCII and the positions for proton loss (represented by red circles) occurring at the ends of the radical cations of these carotenoids. Proton loss would be facilitated by proton acceptors in nearby thylakoid lumen and stroma. Figure 6.1(b) shows the LHCII crystal structure with carotenoids in the same positions as in Figure 6.1(a). Zeaxanthin, which

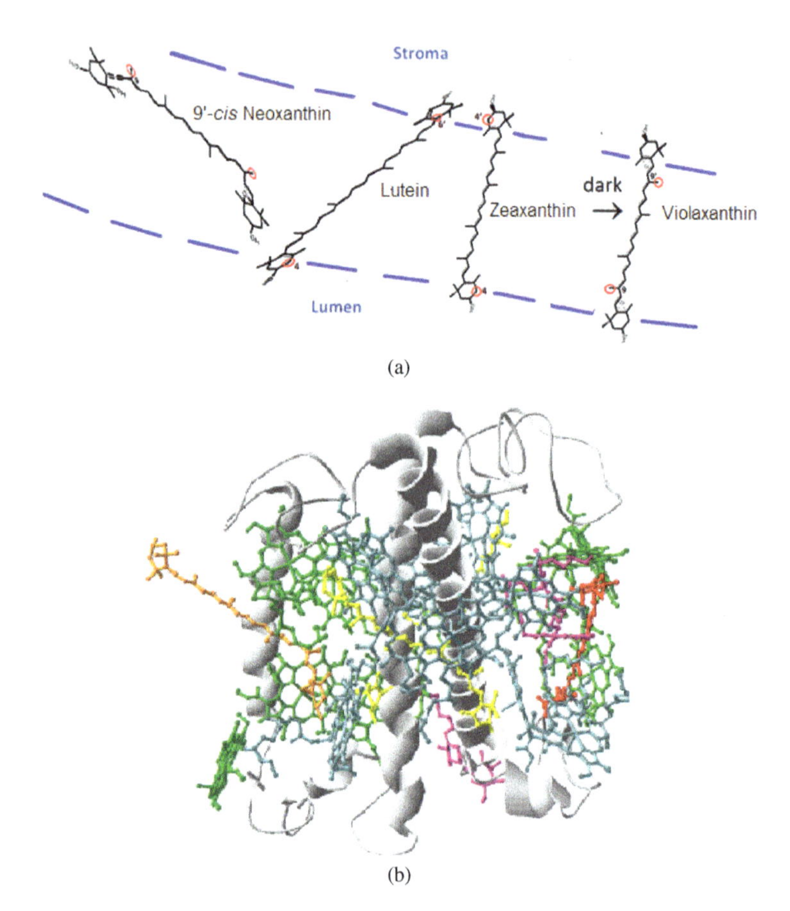

Figure 6.1. (a) Schematic of the position of carotenoids in LHCII. Adapted with permission from Figure 3 of Reference 8. (b) Crystal structure of LHCII. Adapted with permission from Figure 1 of Reference 3.

is accumulated under sunlight and represented in red in Figure 6.1(b), converts to violaxanthin under low light/dark. The lutein molecules found in a cross position are depicted in yellow, and 9′-*cis*-neoxanthin is depicted in orange. For zeaxanthin and lutein spanning across the hydrophobic area and with terminal ends oriented toward stroma and lumen, proton loss is possible at both ends of the radical cations. Carotenoid 9′-*cis*-neoxanthin is buried in the non-polar environment and its epoxy group on one end and the allene bond at the opposite end prevent proton loss at the terminal C5 and C5′-methyl groups by reducing the conjugation crucial for the neutral radical stability. Violaxanthin also contains epoxy groups at both ends which prevents proton loss from positions on the terminal rings of the radical cation to form the neutral radicals.[1] It was noted that the ability to form neutral radicals by deprotonation of the radical cation is correlated with the quenching/non-quenching properties of the four carotenoids present in LHCII.[3] More exactly, zeaxanthin and lutein that are known to be quenchers (and their radical cations have been detected in minor LHCII complexes) can form neutral radicals by proton loss from the radical cation, while violaxanthin and 9′-*cis*-neoxanthin (whose radical cations have not been detected) which are not known to be quenchers do not form neutral radicals by proton loss from the terminal rings.[3]

We have concluded that zeaxanthin and lutein radical cations′ ability to deprotonate and form neutral radicals could be linked to their quenching activity. An additional quenching mechanism involving neutral radicals of zeaxanthin and lutein was thus proposed. For zeaxanthin, for example, a charge transfer complex ($Zea^{\bullet+}\ldots Chl^{\bullet-}$) with 200 ps lifetime is known to be formed (and the radical cation was detected) in the presence of intense sunlight. Deprotonation of the radical cation would form the neutral radical according to $Zea^{\bullet+} \leftrightarrow {}^{\#}Zea^{\bullet} + H^{+}$. A neutral radical like ${}^{\#}Zea^{\bullet}$ would be able to quench the excited state from a neighboring chlorophyll Chl^{*}. Quenching by free radicals is important in liquids and solids and forms the basis for fluorescence detection of reactive oxygen species. Quenching of fluorescence by J exchange for either an excited single or triplet state has been accomplished by attaching a stable nitroxide neutral radical as far away as 9 Å from a fluorescing molecule. In Figure 6.2 is shown the crystal structure of the minor LHC CP29 (Protein Data Bank 3PL9) where the xanthophyll is located between adjacent chlorophylls. One end of the xanthophyll is next to a chlorophyll permitting a charge

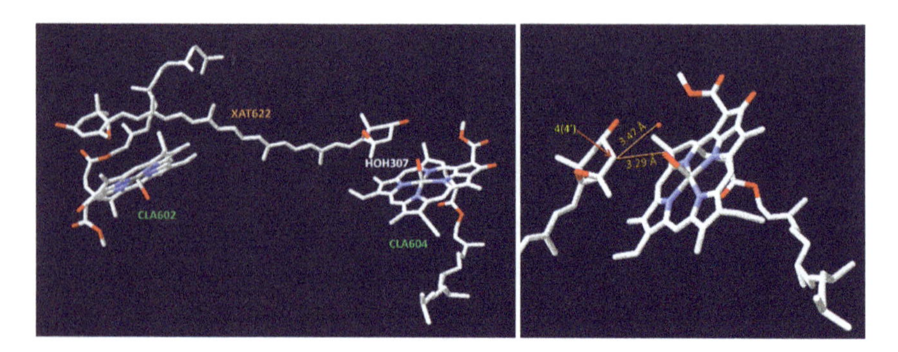

Figure 6.2. Xanthophyll in close proximity (less than 9 Å) to a water molecule. From the structure of minor LHC CP29 (Protein Data Bank 3PL9). Adapted with permission from Figure 4 of Reference 7.

transfer complex to be formed upon light exposure. The negative charge of chlorophyll migrates over adjacent chlorophyll in the 200 ps period, during which time the radical cation formed can transfer a proton to the water molecule located within 9 Å thus forming a quenching neutral radical. This neutral radical species quenches excited chlorophyll during the 200 ps lifetime of the charge transfer complex. The crystal structure of CP29 (Protein Data Bank 3PL9) also shows a lutein molecule with Chl molecules and water molecules situated in close proximity (less than 9 Å) to both ends of the carotenoid.

Zeaxanthin, which accumulates in the presence of light, is a very special and necessary carotenoid. It has a relatively low oxidation potential compared to other carotenoids and thus it can be easily oxidized to form $Zea^{\bullet+}$. A molecule with a higher oxidation potential like astaxanthin, for example, would not form a radical cation as easily and would favor scavenging of radicals like $^{\bullet}OH$, $^{\bullet}OOH$ instead.[11] The radical cation $Zea^{\bullet+}$ is a weak acid (pK_a ~ 4) and deprotonation would easily happen in the presence of proton acceptors like water molecules. Its deprotonation occurs at the C4 or C4′ terminal end to form the neutral radical. $^{\#}Zea^{\bullet}$ neutral radical is long lived and reprotonation is difficult, given the unfavorable potentials. Terminal OH groups of zeaxanthin also facilitate H-aggregation in solvents containing water, causing a 100 nm blue shift.[12] This aggregation reduces reaction with free radicals like $^{\bullet}OH$, $^{\bullet}OOH$ and makes zeaxanthin prone to oxidation by Fe^{3+} or other oxidants to form zeaxanthin radical cation, $Zea^{\bullet+}$. The oxidation potential of zeaxanthin was determined to be to be 571 ± 11 mV, essentially identical to that of β-carotene (567 ± 4 mV) by Niedzwiedzki *et al.*[13] We have also found a relatively low value for the oxidation potential of β-carotene 634 ± 4 mV vs. SCE when compared to other carotenoids, which explained its pro-oxidant effect and formation of the radical cation in reaction with Fe^{3+}.[14] We have also determined a similar oxidation potential for zeaxanthin of 616 mV vs. SCE. The oxidation potential of violaxanthin, 681 ± 14 mV, was found to be higher than that of zeaxanthin, implying that the natural selection of zeaxanthin for its photoprotective role over violaxanthin is owed, at least in part, to its ability to be more easily oxidized to form $Zea^{\bullet+}$.

A study of *Arabidopsis thaliana* as a function of varying light intensity has demonstrated that zeaxanthin neutral radical can form in an organized assembly when $Zea^{\bullet+}...Chl^{\bullet-}$ was generated in bright sunlight.[6] Plants have several redundant mechanisms for protection from excess light, collectively known as non-photochemical quenching (NPQ). One important form of NPQ, known as qE, is characterized by a decrease in fluorescence. *A. thaliana* mutants show a strong connection between zeaxanthin and the qE component of NPQ. The charge transfer complex between Zea and Chl in LHC proteins CP24, CP26 and CP29 in qE-proficient plants was detected by femtosecond transient absorption spectroscopy.

$$Zea + Chl + h\nu \longrightarrow Zea^{\bullet+}...Chl^{\bullet-}$$

Excited Chl generated in these proteins are quenched by formation of the charge transfer complex, producing the radical cation of Zea and the Chl radical anion which then recombine to the ground state in 200 ps. While this is fast, the time scale of proton transfer is measured in femtoseconds when proton donors and acceptors are preassembled. At low light levels ($\leq$5 μE m^{-2} s^{-1}), Zea in the xanthophyll cycle is converted into violaxanthin Vio (which does not support qE) by violaxanthin de-epoxidase. This is triggered by excess protons and a low pH as the photosynthetic apparatus needs every photon available and qE becomes detrimental to the plant. In intense sunlight ($\geq$300 μE m^{-2} s^{-1}), Vio is converted into Zea which forms the charge transfer complex $Zea^{•+}$...$Chl^{•-}$. As discussed above, our electrochemistry studies showed that the carotenoid radical cations are weak acids ($pK_a \sim$ 4–7), and our DFT calculations showed that the most acidic protons for $Zea^{•+}$ occur from the terminal rings at the C4(4′) methylene or from the methyl groups at the C5(5′) carbon, with the C4(4′) being the most favorable. On the other hand, the epoxide groups on the terminal rings of $Vio^{•+}$ prevent proton loss from the terminal rings. The only favorable proton loss for $Vio^{•+}$ is from the methyl protons located on the polyene chain which is not near the acidic stroma/lumen area. Proton loss from $Zea^{•+}$ during qE was examined via H/D exchange using electrospray ionization mass spectrometry in *A. thaliana* plants.[6] Leaves from *A. thaliana* were saturated in D_2O and exposed to levels of sunlight either below or above the threshold to activate qE. When $Zea^{•+}$ deprotonates to form $^{\#}Zea^{•}$, it would eventually gain a deuteron and cause an isotopic shift in the mass spectrum. H/D exchange was observed only for those plants in the presence of sunlight intense enough to trigger qE (>300 μE m^{-2} s^{-1}). These results demonstrate that $Zea^{•+}$ deprotonated as predicted to form $^{\#}Zea^{•}$ during qE. The crystal structure of the qE-active protein CP29 (Protein Data Bank 3 PL9) in Figure 6.2 shows proton acceptors in close proximity to the most acidic carbon C4(4′). The presence of free radicals has been shown to be a potent quencher of excited states by electron exchange-induced quenching. For example, quenching by free radicals is the basis for fluorescence detection of reactive oxygen species. Deprotonation would leave $Chl^{•-}$ and $^{\#}Zea^{•}$ in the protein.

Recombination is prevented because deprotonation drastically shifts the redox potential of Zea. The radical ions of Chl undergo facile electron transfer with other Chl in photosystems I and II as well as in LHCs. The lifetime of $Zea^{\bullet+}$...$Chl^{\bullet-}$ was measured to be about 200 ps. $Chl^{\bullet-}$ charge migration uses the delay in the collapse of $Zea^{\bullet+}$...$Chl^{\bullet}$, allowing a small amount of $Zea^{\bullet+}$ to deprotonate and ready to quench other excited Chl^{*} formed upon continuous light irradiation. Reprotonation of $^{\#}Zea^{\bullet}$ is electrochemically unfavorable, so lifetime of the neutral radical could increase to ms or longer and exist after collapse of the $Zea^{\bullet+}$...$Chl^{\bullet-}$ complex. As mentioned above, an organized assembly can trigger deprotonation of the weak acid radical cation at the terminal C4(4′) positions giving rise to a long lifetime of neutral radical $^{\#}Zea^{\bullet}$ because reprotonation is unfavorable. In the solid state, $^{\#}Zea^{\bullet}$ forms from $Zea^{\bullet+}$ very efficiently as discussed in our previous chapter. The crystal structure of CP29 shows a network of water molecules that lie within hydrogen bonding distances of each other as well as the hydroxyl at C3(3′). This network suggests water plays a functional or structural role in the protein. A model of the experimental procedure and proton transfer mechanism is given in Figure 6.3.

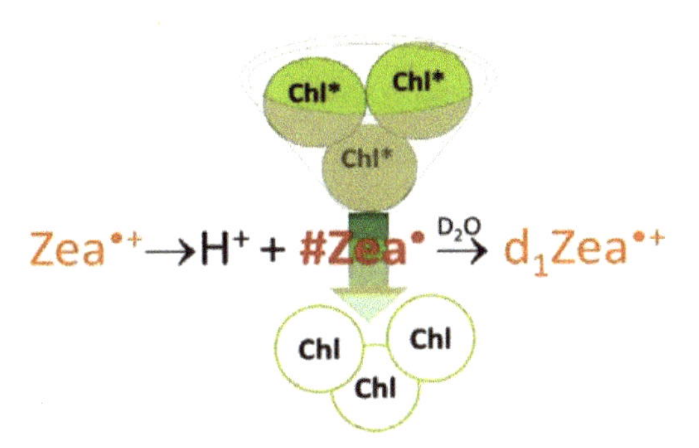

Figure 6.3. Model of experimental procedure and reaction mechanism demonstrating the quenching ability of the zeaxanthin neutral radical $^{\#}Zea^{\bullet}$, as well as the experimental procedure using isotopic labeling detected via liquid chromatography/mass spectrometry to examine its presence. Adapted with permission from Reference 6.

All photosynthetic organisms including plants, algae, and cyanobacteria synthesize carotenoids *de novo* in all kinds of plastids. The primary carotenoids synthesized from lycopene are α-carotene and β-carotene, from which lutein and zeaxanthin, violaxanthin and neoxanthin, are enzymatically formed, respectively. We have shown that the β-carotene radical cation is also easily deprotonated in an artificial system, and the deprotonation is facilitated in the presence of water as proton acceptors. Deprotonation of the β-carotene radical cation to form neutral radicals was also postulated to occur *in vivo* in photosystem II samples.[4] β-carotene neutral radicals were detected optically in photosystem II, explaining the origin of the previously unassigned near-infrared absorption peak at 750 nm and suggesting that the extensive secondary electron transfer pathway in photosystem II may also be involved in proton transfer. A crystal structure showed an electron transfer pathway from a β-carotene molecule with an adjacent proton acceptor to $P680^{\bullet+}$.[4]

Under stress conditions such as high light exposure, nutrient starvation, change in oxygen partial pressure, and high or low temperatures, microalgal metabolism is altered and photosynthetic activity may be reduced. To prevent damage from photo-oxidation, algae can synthesize large amounts of carotenoids that act as antioxidants to fight dangerous free radicals. The antioxidant astaxanthin, a keto carotenoid, is known to accumulate in *Haematococcus pluvialis* green micro algae under unfavorable environmental conditions such as high light exposure, absence of nutrients or in the presence of salt. Our study[9] suggests that astaxanthin's efficiency as a protective agent under high illumination in the presence of metal ions could be related to its ability to form neutral radicals which can be very effective quenchers of the excited states of chlorophyll. Astaxanthin possesses other properties that could aid in fighting under stressful conditions such as its ability to form chelate complexes with metals and to be esterified, its inability to aggregate in the ester form and its high oxidation potential.[9]

References

1. Focsan, A. L., Bowman, M. K., Konovalova, T. A., Molnár, P., Deli, J., Dixon, D. A., & Kispert, L. D. (2008). Pulsed EPR and DFT characterization of radicals produced by photo-oxidation of zeaxanthin and violaxanthin on silica-alumina. *The Journal of Physical Chemistry B, 112*(6), 1806–1819. https://doi.org/10.1021/jp0765650
2. Focsan, A. L. (2008). EPR and DFT studies of proton loss from carotenoid radical cations. PhD Dissertation, The University of Alabama, Tuscaloosa.
3. Focsan, A. L., Molnár, P., Deli, J., & Kispert, L. (2009). Structure and properties of 9′-*cis* neoxanthin carotenoid radicals by electron paramagnetic resonance measurements and density functional theory calculations: Present in LHC II? *The Journal of Physical Chemistry B, 113*(17), 6087–6096. https://doi.org/10.1021/jp810604s
4. Gao, Y., Shinopoulos, K. E., Tracewell, C. A., Focsan, A. L., Brudvig, G. W., & Kispert, L. D. (2009). Formation of carotenoid neutral radicals in photosystem II. *The Journal of Physical Chemistry B, 113*(29), 9901–9908. https://doi.org/10.1021/jp8075832
5. Focsan, A. L., Bowman, M. K., Molnár, P., Deli, J., & Kispert, L. D. (2011). Carotenoid radical formation: Dependence on conjugation length. *The Journal of Physical Chemistry B, 115*(30), 9495–9506. https://doi.org/10.1021/jp204787b
6. Magyar, A., Bowman, M. K., Molnár, P., & Kispert, L. (2013). Neutral carotenoid radicals in photoprotection of wild-type *Arabidopsis thaliana*. *The Journal of Physical Chemistry B, 117*(8), 2239–2246. https://doi.org/10.1021/jp306387e
7. Focsan, A., L., Magyar, A., & Kispert, L. D. (2015). Chemistry of carotenoid neutral radicals. *Archives of Biochemistry and Biophysics, 572*, 167–174. https://doi.org/10.1016/j.abb.2015.02.005
8. Focsan, A. L., & Kispert, L. D. (2017). Radicals formed from proton loss of carotenoid radical cations: A special form of carotenoid neutral radical

occurring in photoprotection. *Journal of Photochemistry and Photobiology B: Biology, 166*, 148–157. https://doi.org/10.1016/j.jphotobiol.2016.11.015

9. Focsan, A. L., Polyakov, N. E., & Kispert, L. D. (2017). Photo protection of *Haematococcus pluvialis* algae by astaxanthin: Unique properties of astaxanthin deduced by EPR, optical and electrochemical studies. *Antioxidants, 6*(4), 80. https://doi.org/10.3390/antiox6040080
10. Focsan, A. L., Polyakov, N. E., & Kispert, L. D. (2021). Carotenoids: Importance in daily life-Insight gained from EPR and ENDOR. *Applied Magnetic Resonance, 52*(8), 1093–1112. https://doi.org/10.1007/s00723-021-01311-8
11. Polyakov, N. E., Kruppa, A. I., Leshina, T. V., Konovalova, T. A., & Kispert, L. D. (2001). Carotenoids as antioxidants: Spin trapping EPR and optical study. *Free Radical Biology and Medicine, 31*(1), 43–52. https://doi.org/10.1016/s0891-5849(01)00547-0
12. Polyakov, N. E., Magyar, A., & Kispert, L. D. (2013). Photochemical and optical properties of water-soluble xanthophyll antioxidants: Aggregation vs complexation. *The Journal of Physical Chemistry B, 117*(35), 10173–10182. https://doi.org/10.1021/jp4062708
13. Niedzwiedzki, D., Rusling J. F., & Frank, H. A. (2005). Voltammetric redox potentials of carotenoids associated with the xanthophyll cycle in photosynthesis. *Chemical Physics Letters, 415*, 308–312. https://doi.org/10.1016/j.cplett.2005.09.010
14. Polyakov, N. E., Leshina, T. V., Konovalova, T. A., & Kispert, L. D. (2001). Carotenoids as scavengers of free radicals in a Fenton reaction: Antioxidants or pro-oxidants? *Free Radical Biology and Medicine, 31*(3), 398–404. https://doi.org/10.1016/s0891-5849(01)00598-6

7

Carotenoid Analysis Sources

Due to their fundamental roles in various biological processes, carotenoids can be described as "molecules essential for life". These organic pigments are essential for both plants and many organisms including humans, playing critical roles in photosynthesis, protection against oxidative damage, and various physiological functions. These multifaceted molecules contribute significantly to the vitality and sustainability of life forms across different kingdoms. Exploring carotenoids enriches our understanding of life itself. They deserve and demand our scientific curiosity.

We have shown that different electron paramagnetic resonance (EPR) techniques, UV-vis spectroscopy, cyclic voltammetry, or a combination of techniques including density functional theory can provide a more comprehensive understanding of carotenoids and their complex interactions.

Other analytical techniques used to identify and quantify carotenoids not discussed in this book are high-performance liquid chromatography (HPLC), liquid chromatography-mass spectrometry, nuclear magnetic resonance (NMR), Fourier transform infrared spectroscopy, ultrafast transient absorption spectroscopy of photoregulation, etc. Also, using computational methods, scientists can calculate the electronic structure, geometries, potential energies and couplings in carotenoids to simulate spectra, which can be compared with experimental data to validate the results. The primary literature about natural occurrence and isolation, and spectroscopic data for identification, can be found in the book: Britton, G., Liaaen-Jensen, S., & Pfander, H. (2008). *Carotenoids Handbook*. Birkhäuser Verlag.

The carotenoid field originates from plant natural product chemistry. However, in recent times, the use of carotenoids as food pigments and vital nutrients in human diet has attracted a surge of newcomers to the field, with more focus on nutrition and health. Consequently, carotenoid research has expanded into diverse realms, and the development of these research fields prompted the creation of a focused organization, the International Carotenoid Society (www.carotenoidsociety.org), whose main objective is to serve as a central hub and platform for all matters pertaining to carotenoids. By joining the International Carotenoid Society, one would gain access to the latest carotenoid research from various meetings, and other means for facilitating research in this field.

Here we provide some of the key resources to study carotenoids similarly to our studies.

Electrochemistry products can be found at BASi Research Products and Gamry Instruments. This includes electrochemical analyzers and other instrumentation as well as guaranteed high-quality, practical accessories including a large selection of electrodes, accessories and complete cell packages. The Digisim program that we used in our simulation has been discontinued. Another alternative could be KISSA-1D Electrochemistry Simulation Software or DigiElch 8 from ElchSoft sold by Gamry Instruments.

https://www.basinc.com/electrochemistry-simulation-software

https://www.gamry.com/digielch-electrochemical-simulation-software/

There are a multitude of extraction methods for carotenoids from natural sources. A table with references is provided at the end of this chapter (Table 7.1). For performing HPLC a special carotenoid-specific column is sold by YMC. The carotenoid C30 stationary phase provides sufficient phase thickness to enhance interaction with long-chain molecules, and also geometric and positional isomers of conjugated double-bonding systems are recognized and resolved.

https://www.ymc.co.jp/en/columns/ymc_carotenoid/ or https://www.ymcamerica.com/ymc-carotenoid-c30/

The National High Magnetic Field Laboratory (NHMFL) is a research facility located at Florida State University, the University of Florida, and Los Alamos National Laboratory. The laboratory is funded by the National Science Foundation and the State of Florida. It provides researchers with a wide range of high magnetic field instruments capable of generating magnetic fields up to the world-record 45 tesla to be used in a variety of fields, including physics, chemistry, materials science, and biology. One of the primary research areas at NHMFL is high magnetic field physics, which includes the study of materials, condensed matter, and biological systems under extreme magnetic fields. NHMFL also conducts research in other areas, including magnetic resonance imaging, NMR, EPR, ion cyclotron resonance (ICR) mass spectrometry or high-temperature superconductivity. The laboratory offers a variety of training and education programs for students and researchers.

1800 E. Paul Dirac Drive Tallahassee, FL 32310

EPR: https://nationalmaglab.org/user-facilities/emr/

NMR: https://nationalmaglab.org/user-facilities/nmr-mri

ICR: https://nationalmaglab.org/user-facilities

The University of Alabama's Department of Chemistry and Biochemistry offers a variety of research instrumentation for its faculty, staff, and students. The department has a NMR facility, a mass spectrometry facility, an EPR facility, access to Alabama Analytical Research Center and access to surface characterization instruments.

2004 Shelby Hall

(205) 348-5954

chemistry@as.ua.edu

Mailing Address

Box 870336

The University of Alabama

Tuscaloosa, AL 35487-0336

https://chemistry.ua.edu/department-overview/facilities/research-instrumentation/

This department is one of the few in the nation with its own glass-blowing shop, where a professional glassblower fashions custom lab equipment for faculty research. This was useful in our research for designing special cells or apparatus.

https://as.ua.edu/research/facilities/glassblowing-facility/

Northwestern

INSTITUTE FOR
SUSTAINABILITY AND ENERGY

The Institute for Sustainability and Energy at Northwestern University (ISEN) includes the Center for Light Energy Activated Redox Processes, Solar Fuels Institute, Center for Catalysis and Surface Science, and Center for Advanced Materials for Energy and Environment. Northwestern University's Office for Research core mission-driven research centers focused on sustainability and energy provide a wide range of instrumentation and services and also partner with Argonne National Laboratory. The Chemistry department at Northwestern University is affiliated with ISEN. Professor Wasielewski's research lab at Northwestern University centers on light-driven charge transfer and transport in molecules and materials, photosynthesis, nanoscale materials for solar energy conversion, spin dynamics of multi-spin molecules, quantum information science, and time-resolved optical and electron paramagnetic resonance spectroscopy. Special techniques used include steady-state absorption and emission, transient absorption spectroscopy, transient absorption microscopy, two-dimensional electronic spectroscopy, time-resolved femtosecond stimulated Raman spectroscopy, time-resolved electron paramagnetic resonance spectroscopy, pulsed electron magnetic resonance and optically detected magnetic resonance spectroscopies.

Institute for Sustainability and Energy at Northwestern (ISEN)
2145 Sheridan Rd.
Evanston, IL 60208, US
https://isen.northwestern.edu/

Department of Chemistry, Prof. Michael Wasielewski
2145 Sheridan Road
Northwestern University
Evanston, IL 60208-3113
https://sites.northwestern.edu/wasielewski/research/

The Environmental Molecular Sciences Laboratory (EMSL) is a U.S. Department of Energy national scientific user facility located at the Pacific Northwest National Laboratory in Richland, Washington. EMSL is a multidisciplinary research center that focuses on the study of complex biological, environmental, and energy-related systems. The laboratory provides a suite of advanced instrumentation, including EPR, mass spectrometry, NMR spectroscopy, X-ray diffraction, electron microscopy, and computational resources to support cutting-edge research in a variety of fields. The center's state-of-the-art instrumentation, technical support, and training opportunities make it a valuable resource for researchers across many fields.

3335 Innovation Boulevard, Richland, WA 99354

509.371.6003

emsl@pnnl.gov

https://www.emsl.pnnl.gov/science/instruments-resources

The Advanced Center for Electron Spin Resonance Technology (ACERT) is a national resource center for modern electron spin resonance (ESR) spectroscopy located at Cornell University's Department of Chemistry and Chemical Biology in Ithaca, New York. The center is funded by the National Institutes of Health (NIH) and provides access to state-of-the-art ESR instrumentation and expertise to researchers in academia, government, and industry. The new National Biomedical Resource for Advanced ESR Spectroscopy which kept the original acronym, ACERT, places greater emphasis on service to the national ESR community. The center provides technical support and training to users, helping them design experiments, interpret data, and troubleshoot instrument issues.

Dept. of Chemistry & Chemical Biology
259 Feeney Way Cornell University Ithaca, NY 14853, U.S.A.
https://www.acert.cornell.edu/

The National Biomedical EPR Center is a research facility dedicated to advancing the field of EPR spectroscopy in biomedical research with a focus on free radicals, spin labeling, metal complexes, and metalloproteins. The center is located at the Medical College of Wisconsin in Milwaukee, Wisconsin, USA. This US Research Resource was supported by NIH from 1976 to 2019 and it has positioned itself to maintain its status as an EPR resource to the scientific community.

Department of Biophysics
Medical College of Wisconsin
8701 Watertown Plank Rd.
Milwaukee, WI 53226
https://www.mcw.edu/departments/national-biomedical-epr-center

Table 7.1. Different extraction methods of carotenoids from natural sources.

Low-pressure extraction (conventional solvent extraction)	1. Shi, J., Yi, C., Ye, X., Xue, S., Jiang, Y. Ma, Y., & Liu, D. (2010). Effects of supercritical CO_2 fluid parameters on chemical composition and yield of carotenoids extracted from pumpkin. *LWT — Food Science and Technology, 43*(1), 39–44. https://doi.org/10.1016/j.lwt.2009.07.003 2. Choudhari, S. M., & Singhal, R. S. (2008). Supercritical carbon dioxide extraction of lycopene from mated cultures of Blakeslea trispora NRRL 2895 and 2896. *Journal of Food Engineering, 89*(3), 349–354. https://doi.org/10.1016/j.jfoodeng.2008.05.016 3. Shi, X., Wu, H., Shi, J., Xue, S.J., Wang, D., Wang, W., Cheng, A., Gong, Z., Chen, X., & Wang, C. (2013). Effect of modifier on the composition and antioxidant activity of carotenoid extracts from pumpkin (Cucurbita maxima) by supercritical CO_2. *LWT — Food Science and Technology, 51*(2), 433–440. http://dx.doi.org/10.1016/j.lwt.2012.11.003 4. Kim, S. M., Kang, S.-W., Kwon, O.-N., Chung, D., Pan, C.-H. (2012). Fucoxanthin as a major carotenoid in Isochrysis aff. galbana: Characterization of extraction for commercial application. *Journal of the Korean Society for Applied Biological Chemistry, 55*, 477–483. https://doi.org/10.1007/s13765-012-2108-3 5. Silveira, S. T., Burkert, J. F., Costa, J. A., Burkert, C. A., & Kalil, S. J. (2007). Optimization of phycocyanin extraction from Spirulina platensis using factorial design. *Bioresource Technology, 98*(8), 1629–1634. https://doi.org/10.1016/j.biortech.2006.05.050
Soxhlet extraction	1. Mezzomo, N., Maestri, B., dos Santos, R. L., Maraschin, M., & Ferreira, S. R. (2011). Pink shrimp (P. brasiliensis and P. paulensis) residue: influence of extraction method on carotenoid concentration. *Talanta, 85*(3), 1383–1391. https://doi.org/10.1016/j.talanta.2011.06.018 2. Saldana, M. D. A., Sun, L., Guigard, S. E., & Temelli, F. (2006). Comparison of the solubility of β-carotene in supercritical CO_2 based on a binary and a multicomponent complex system. *The Journal of Supercritical Fluids, 37*(3), 342–349. https://doi.org/10.1016/j.supflu.2006.01.010

	3. Bashipour, F., & Ghoreishi, S. M. (2012). Experimental optimization of supercritical extraction of β-carotene from Aloe barbadensis Miller via genetic algorithm. *The Journal of Supercritical Fluids, 72*, 312–319. https://doi.org/10.1016/j.supflu.2012.10.005
	4. Cardenas-Toro, F. P., Alcázar-Alay, S. C., Coutinho, J. P., Godoy, H. T., Forster-Carneiro, T., & Meireles, M. A. A. (2015). Pressurized liquid extraction and low-pressure solvent extraction of carotenoids from pressed palm fiber: Experimental and economical evaluation. *Food and Bioproducts Processing, 94*, 90–100. https://doi.org/10.1016/j.fbp.2015.01.006
High shear dispergator extraction	1. Tiwari, S., Upadhyay, N., Singh, A. K., Meena, G. S., & Arora, S. Organic solvent-free extraction of carotenoids from carrot bio-waste and its physico-chemical properties. *Journal of Food Science and Technology, 56*(10), 4678–4687. https://doi.org/10.1007/s13197-019-03920-5
	2. Baria, B., Upadhyay, N., Singh, A. K., & Malhotra, R. K. (2019). Optimization of 'green' extraction of carotenoids from mango pulp using split plot design and its characterization. *LWT — Food Science and Technology, 104*, 186–194. https://doi.org/10.1016/j.lwt.2019.01.04
High-pressure homogenization	1. Tavanandi, H. A., Vanjari, P., & Raghavarao, K. (2019). Synergistic method for extraction of high purity Allophycocyanin from dry biomass of Arthrospira platensis and utilization of spent biomass for recovery of carotenoids. *Separation and Purification Technology, 225*, 97–111. https://doi.org/10.1016/j.seppur.2019.05.064.
	2. Spiden, E. M., Yap, B. H., Hill, D. R., Kentish, S. E., Scales, P. J., & Martin, G. J. (2013). Quantitative evaluation of the ease of rupture of industrially promising microalgae by high pressure homogenization. *Bioresource Technology, 140*, 165–171. https://doi.org/10.1016/j.biortech.2013.04.074
	3. Bernaerts, T. M., Verstreken, H., Dejonghe, C., Gheysen, L., Foubert, I., Grauwet, T., & Van Loey, A. M. (2020). Cell disruption of Nannochloropsis sp. improves in vitro bioaccessibility of carotenoids and ω3-LC-PUFA. *Journal of Functional Foods, 65*, 103770. https://doi.org/10.1016/j.jff.2019.103770

(*Continued*)

Table 7.1. (*Continued*)

Continuous pressurized solvent extraction	1. Amaro, H. M., Guedes, A. C., Preto, M. A., Sousa-Pinto, I., & Malcata, F. X. (2018). *Gloeothece* sp. as a nutraceutical source — An improved method of extraction of carotenoids and fatty acids. *Marine Drugs, 16*(9), 327. https://doi.org/10.3390/md16090327
Pressurized liquid extraction	1. Jaime, L., Guez-Meizoso, I. R., Cifuentes, A., Santoyo, S. Suarez, S., Ibanez, E., & Senorans, F. J. (2010). Pressurized liquids as an alternative process to antioxidant carotenoids' extraction from Haematococcus pluvialis microalgae. *LWT — Food Science and Technology, 43*(1), 105–112. https://doi.org/10.1016/j.lwt.2009.06.023 2. Cardenas-Toro, F. P., Alcázar-Alay, S. C., Coutinho, J. P., Godoy, H. T., Forster-Carneiro, T., & Meireles, M. A. A. (2015). Pressurized liquid extraction and low-pressure solvent extraction of carotenoids from pressed palm fiber: Experimental and economical evaluation. *Food and Bioproducts Processing, 94*, 90–100. https://doi.org/10.1016/j.fbp.2015.01.006 3. Tumbas Šaponjac, V., Kovačević, S., Šeregelj, V., Šovljanski, O., Mandić, A., Ćetković, G., Vulić, J., Podunavac-Kuzmanović, S., & Čanadanović-Brunet, J. (2021). Improvement of carrot accelerated solvent extraction efficacy using experimental design and chemometric techniques. *Processes*, *9*(9), 1652. https://doi.org/10.3390/pr9091652 4. Cha, K. H., Lee, H. J., Koo, S. Y., Song, D. G., Lee, D. U., & Pan, C. H. (2010). Optimization of pressurized liquid extraction of carotenoids and chlorophylls from Chlorella vulgaris. *Journal of Agricultural and Food Chemistry, 58*(2), 793–797. https://doi.org/10.1021/jf902628j 5. Rodríguez-Meizoso, I., Jaime, L., Santoyo, S., Cifuentes, A., García-Blairsy Reina, G., Señoráns, F. J., & Ibáñez, E. (2008). Pressurized fluid extraction of bioactive compounds from Phormidium species. *Journal of Agricultural and Food Chemistry*, *56*(10), 3517–3523. https://doi.org/10.1021/jf703719p

Microwave-assisted extraction	1. Sharma, M., & Bhat, R. (2021). Extraction of carotenoids from pumpkin peel and pulp: Comparison between innovative green extraction technologies (ultrasonic and microwave-assisted extractions using corn oil). *Foods, 10*(4), 787. https://doi.org/10.3390/foods10040787 2. Hiranvarachat, B., Devahastin, S., Chiewchan, N., & Raghavan, G. S.V. (2013). Structural modification by different pretreatment methods to enhance microwave-assisted extraction of β-carotene from carrots. *Journal of Food Engineering, 115*(2), 190–197. https://doi.org/10.1016/j.jfoodeng.2012.10.012 3. Ho, K. K. H. Y., Ferruzzi, M. G., Liceaga, A. M., & San Martin-Gonzalez, M. F. (2015). Microwave-assisted extraction of lycopene in tomato peels: Effect of extraction conditions on all-trans and cis-isomer yields. *LWT — Food Science and Technology, 62*(1), 160–168. https://doi.org/10.1016/j.lwt.2014.12.061 4. Chutia, H., & Mahanta, C. L. (2021). Green ultrasound and microwave extraction of carotenoids from passion fruit peel using vegetable oils as a solvent: Optimization, comparison, kinetics, and thermodynamic studies. *Innovative Food Science & Emerging Technologies, 67,* 102547. https://doi.org/10.1016/j.ifset.2020.102547 5. Baria, B., Upadhyay, N., Singh, A. K., & Malhotra, R. K. (2019). Optimization of 'green' extraction of carotenoids from mango pulp using split plot design and its characterization. *LWT — Food Science and Technology, 104,* 186–194. https://doi.org/10.1016/j.lwt.2019.01.04 6. Elik, A., Yanık, D. K., & Göğüş, F. (2020). Microwave-assisted extraction of carotenoids from carrot juice processing waste using flaxseed oil as a solvent. *Lebensmittel-Wissenschaft + Technologie, 123,* 109100. https://doi.org/10.1016/j.lwt.2020.109100 7. Sharma, M., Hussain, S., Shalima, T., Aav, R., & Bhat, R. (2022). Industrial crops & products valorization of seabuckthorn pomace to obtain bioactive carotenoids: An innovative approach of using green extraction techniques (ultrasonic and microwave-Assisted extractions) synergized with green solvents (Edible oils). *Industrial Crops and Products, 175,* 14257. https://doi.org/10.1016/j.indcrop.2021.114257

(*Continued*)

Table 7.1. (*Continued*)

	8. Juin, C., Chérouvrier, J.-R., Thiéry, V., Gagez, A.-L., Bérard, J.-B., Joguet, N., Kaas, R., Cadoret, J.-P., & Picot, L. (2015). Microwave-assisted extraction of phycobiliproteins from Porphyridium purpureum. *Biotechnology and Applied Biochemistry, 175,* 1–15. https://doi.org/10.1007/s12010-014-1250-2
	9. Pasquet, V., Chérouvrier, J.-R., Farhat, F., Thiéry, V., Piot, J.-M., Bérard, J.-B., Kaas, R., Serive, B., Patrice, T., & Cadoret, J.-P. (2011). Study on the microalgal pigments extraction process: Performance of microwave assisted extraction. *Process Biochemistry, 46,* 59–67. https://doi.org/10.1016/j.procbio.2010.07.009
	10. Hiranvarachat, B., Devahastin, S., Chiewchan, N., & Vijaya Raghavan, G. S. (2013). Structural modification by different pretreatment methods to enhance microwave-assisted extraction of β-carotene from carrots. *Journal of Food Engineering, 115*(2), 190–197. https://doi.org/10.1016/j.jfoodeng.2012.10.012
	11. Pasquet, V., Chérouvrier, J.-R., Farhat, F., Thiéry, V., Piot, J.-M., Bérard, J.-B., Kaas, R., Serive, B., Patrice, T., Cadoret, J.- P., & Picot, L. (2011). Study on the microbial pigments extraction process: performance of microwave assisted extraction. *Process Biochemistry, 46,* 59–67. http://dx.doi.org/10.1016/j.procbio.2010.07.009
Ultrasound-assisted extraction	1. Sharma, M., & Bhat, R. (2021). Extraction of carotenoids from pumpkin peel and pulp: Comparison between innovative green extraction technologies (ultrasonic and microwave-assisted extractions using corn oil). *Foods, 10*(4), 787. https://doi.org/10.3390/foods10040787
	2. Umair, M., Jabbar, S., Nasiru, M. M., Lu, Z., Zhang, J., Abid, M., Murtaza, M. A., Kieliszek, M., & Zhao, L. (2021). Ultrasound-assisted extraction of carotenoids from carrot pomace and their optimization through response surface methodology. *Molecules, 26*(22), 6763. https://doi.org/10.3390/molecules26226763
	3. Civan, M., & Kumcuoglu, S. (2019). Green ultrasound-assisted extraction of carotenoid and capsaicinoid from the pulp of hot pepper paste based on the bio-refinery concept. *LWT — Food Science and Technology, 113,* 108320. https://doi.org/10.1016/j.lwt.2019.108320

	4. Chutia, H., & Mahanta, C. L. (2021). Green ultrasound and microwave extraction of carotenoids from passion fruit peel using vegetable oils as a solvent: Optimization, comparison, kinetics, and thermodynamic studies. *Innovative Food Science & Emerging Technologies, 67*, 102547. https://doi.org/10.1016/j.ifset.2020.102547 5. Bhimjiyani, V. H., Borugadda, V. B., Naik, S., & Dalai, A. K. (2021). Enrichment of flaxseed (*Linum usitatissimum*) oil with carotenoids of sea buckthorn pomace via ultrasound-assisted extraction technique: Enrichment of flaxseed oil with sea buckthorn. *Current Research in Food Science, 4*, 478–488. https://doi.org/10.1016/j.crfs.2021.07.006 6. Stupar, A., Šeregelj, V., Ribeiro, B. D., Pezo, L., Cvetanović, A., Mišan, A., & Marrucho, I. (2021). Recovery of β-carotene from pumpkin using switchable natural deep eutectic solvents. *Ultrasonics Sonochemistry, 76*, 105638. https://doi.org/10.1016/j.ultsonch.2021.105638 7. Dey, S., & Rathod, V. K. (2013). Ultrasound assisted extraction of β-carotene from Spirulina platensis. *Ultrasonics Sonochemistry, 20*(1), 271–276. https://doi.org/10.1016/j.ultsonch.2012.05.010 8. Li, Y., Fabiano-Tixier, A. S., Tomao, V., Cravotto, G., & Chemat, F. (2013). Green ultrasound-assisted extraction of carotenoids based on the bio-refinery concept using sunflower oil as an alternative solvent. *Ultrasonics Sonochemistry, 20*(1), 12–18. https://doi.org/10.1016/j.ultsonch.2012.07.005
Pulsed electric field extraction	1. Siddeeg, A., Faisal Manzoor, M., Haseeb Ahmad, M., Ahmad, N., Ahmed, Z., Kashif Iqbal Khan, M., Aslam Maan, A., Mahr-Un-Nisa, Zeng, X.-A., & Ammar, A.-F. (2019). Pulsed electric field-assisted ethanolic extraction of date palm fruits: Bioactive compounds, antioxidant activity and physicochemical properties. *Processes, 7*(9), 585. https://doi.org/10.3390/pr7090585 2. López-Gámez, G., Elez-Martínez, P., Martín-Belloso, O., & Soliva-Fortuny, R. (2021). Pulsed electric field treatment strategies to increase bioaccessibility of phenolic and carotenoid compounds in oil-added carrot purees. *Food Chemistry, 364*, 130377. https://doi.org/10.1016/j.foodchem.2021.130377

(Continued)

Table 7.1. (*Continued*)

	3. Pataro, G., Carullo, D., Falcone, M., & Ferrari, G. (2020). Recovery of lycopene from industrially derived tomato processing by-products by pulsed electric fields-assisted extraction. *Innovative Food Science & Emerging Technologies, 63*, 102369. https://doi.org/10.1016/j.ifset.2020.102369 4. Chittapun, S., Jonjaroen, V., Khumrangsee, K., & Charoenrat, T. (2020). C-phycocyanin extraction from two freshwater cyanobacteria by freeze thaw and pulsed electric field techniques to improve extraction efficiency and purity. *Algal Research-Biomass Biofuels and Bioproducts, 46*, 101789. https://doi.org/10.1016/j.algal.2020.101789 5. Martínez, J. M., Delso, C., Álvarez, I., & Raso, J. (2019). Pulsed electric field permeabilization and extraction of phycoerythrin from Porphyridium cruentum. *Algal Research, 37*, 51–56. https://doi.org/10.1016/j.algal.2018.11.005 6. Ahirwar, A., Khan, M. J., Sirotiya, V., Mourya, M., Rai, A., Schoefs, B., Marchand, J., Varjani, S., & Vinayak, V. (2023). Pulsed electric field–assisted cell permeabilization of microalgae (*Haematococcus pluvialis*) for milking of value-added compounds. *BioEnergy Research, 16*, 311–324. https://doi.org/10.1007/s12155-022-10414-4
Supercritical fluid extraction	1. Miękus, N., Iqbal, A., Marszałek, K., Puchalski, C., & Świergiel, A. (2019). Green chemistry extractions of carotenoids from Daucus carota L. — Supercritical carbon dioxide and enzyme-assisted methods. *Molecules, 24*(23), 4339. http://dx.doi.org/10.3390/molecules24234339 2. Vafaei, N., Rempel, C. B., Scanlon, M. G., Jones, P. J. H., & Eskin, M. N. A. (2022). Application of supercritical fluid extraction (SFE) of tocopherols and carotenoids (hydrophobic antioxidants) compared to non-SFE methods. *AppliedChem, 2*(2), 68–92. http://dx.doi.org/10.3390/appliedchem2020005 3. Zaghdoudi, K., Framboisier, X., Frochot, C., Vanderesse, R., Barth, D., Kalthoum-Cherif, J., Blanchard, F., & Guiavarc'h (2016). Response surface methodology applied to supercritical fluid extraction (SFE) of carotenoids from Persimmon (Diospyros kaki L.). *Food Chemistry, 208*, 209–219. http://dx.doi.org/10.1016/j.foodchem.2016.03.104

4. Durante, M., Lenucci, M., & Mita, G. (2014). Supercritical carbon dioxide extraction of carotenoids from pumpkin (Cucurbita spp.): A review. *International Journal of Molecular Sciences, 15*(4), 6725–6740. https://doi.org/10.3390/ijms15046725
5. de Andrade Lima, M., Kestekoglou, I., Charalampopoulos, D., & Chatzifragkou, A. (2019). Supercritical fluid extraction of carotenoids from vegetable waste matrices. *Molecules, 24*(3), 466. https://doi.org/10.3390/molecules24030466
6. de Andrade Lima, M., Charalampopoulos, D., & Chatzifragkou, A. (2018). Optimisation and modelling of supercritical CO2 extraction process of carotenoids from carrot peels. *The Journal of Supercritical Fluids, 133*, 94–102. https://doi.org/10.1016/j.supflu.2017.09.028
7. de Andrade Lima, M., Kestekoglou, I., Charalampopoulos, D., & Chatzifragkou, A. (2019). Supercritical fluid extraction of carotenoids from vegetable waste matrices. *Molecules, 24*(3), 466. https://doi.org/10.3390/molecules24030466
8. Jaime, L., Mendiola, J. A., Ibáñez, E., Martin-Alvarez, P. J., Cifuentes, A., Reglero, G., & Señoráns, F. J. (2007). Beta-carotene isomer composition of sub- and supercritical carbon dioxide extracts. Antioxidant activity measurement. *Journal of Agricultural and Food Chemistry, 55*(26), 10585–10590. https://doi.org/10.1021/jf0711789
9. Hosseini, S. R. P., Tavakoli, O., Sarrafzadeh, M. H. (2017). Experimental optimization of SC-CO_2 extraction of carotenoids from Dunaliella salina. *The Journal of Supercritical Fluids, 121*, 89–95. https://doi.org/10.1016/j.supflu.2016.11.006.
10. Liau, B.-C., Shen, C.-T., Liang, F.-P., Hong, S.-E., Hsu, S.-L., Jong, T.-T., & Chang, C.-M. J. (2010). Supercritical fluids extraction and anti-solvent purification of carotenoids from microalgae and associated bioactivity. *The Journal of Supercritical Fluids, 55*, 169–175. https://doi.org/10.1016/j.supflu.2010.07.002

(*Continued*)

Table 7.1. (*Continued*)

	11. Macıas-Sánchez, M., Mantell, C., Rodrıguez, M., de La Ossa, E. M., Lubián, L., & Montero, O. (2005). Supercritical fluid extraction of carotenoids and chlorophyll a from Nannochloropsis gaditana. *Journal of Food Engineering, 66*(2), 245–251. https://doi.org/10.1016/j.jfoodeng.2004.03.021. 12. Guedes, A. C., Gião, M. S., Matias, A. A., Nunes, A. V., Pintado, M. E., Duarte, C. M., & Malcata, F. X. (2013). Supercritical fluid extraction of carotenoids and chlorophylls a, b and c, from a wild strain of Scenedesmus obliquus for use in food processing. *Journal of Food Engineering, 116*(2), 478–482. https://doi.org/10.1016/j.jfoodeng.2012.12.015 13. Sanzo, G. D., Mehariya, S., Martino, M., Larocca, V., Casella, P., Chianese, S., Musmarra, D., Balducchi, R., & Molino, A. (2018). Supercritical carbon dioxide extraction of astaxanthin, lutein, and fatty acids from *Haematococcus pluvialis* microalgae. *Marine Drugs, 16*(9), 334. https://doi.org/10.3390/md16090334 14. Nobre, B., Marcelo, F., Passos, R., Beirão, L., Palavra, A., Gouveia, L., & Mendes, R. (2006). Supercritical carbon dioxide extraction of astaxanthin and other carotenoids from the microalga *Haematococcus pluvialis. European Food Research and Technology, 223*, 787–790. https://doi.org/10.1007/s00217-006-0270-8.u
Enzyme-assisted extraction	1. Zuorro, A., Fidaleo, M., & Lavecchia, R. (2011). Enzyme-assisted extraction of lycopene from tomato processing waste. *Enzyme and Microbial Technology, 49*(6–7), 567–573. https://doi.org/10.1016/j.enzmictec.2011.04.020 2. Barzana, E., Rubio, D., Santamaria, R. I., Garcia-Correa, O., Garcia, F., Ridaura Sanz, V. E., & López-Munguía, A. (2002). Enzyme-mediated solvent extraction of carotenoids from marigold flower (Tagetes erecta). *Journal of Agricultural and Food Chemistry, 50*(16), 4491–4496. https://doi.org/10.1021/jf025550q 3. Ricarte, G. N., Coelho, M. A. Z., Marrucho, I. M., & Ribeiro, B. D. (2020). Enzyme-assisted extraction of carotenoids and phenolic compounds from sunflower wastes using green solvents. *3 Biotech, 10*(9), 405. https://doi.org/10.1007/s13205-020-02393-0

	4. Nath, P., Kaur, C., Rudra, S.G., & Varghese, E. (2016). Enzyme-assisted extraction of carotenoid-rich extract from Red Capsicum (*Capsicum annuum*). *Agricultural Research, 5*, 193–204. https://doi.org/10.1007/s40003-015-0201-7
High-voltage electrical discharge	1. Zhang, R., Marchal, L., Lebovka, N., Vorobiev, E., & Grimi, N. (2020). Two-step procedure for selective recovery of bio-molecules from microalga Nannochloropsis oculata assisted by high voltage electrical discharges. *Bioresource Technology, 302*, 122893. https://doi.org/10.1016/j.biortech.2020.122893
Laser	1. McMillan, J. R., Watson, I.A., Ali, M., & Jaafar, W. (2013). Evaluation and comparison of algal cell disruption methods: Microwave, waterbath, blender, ultrasonic and laser treatment. *Applied Energy, 103*, 128–134. https://doi.org/10.1016/j.apenergy.2012.09.020

Index